predictably irrational

revised
and expanded
edition

predictably irrational

revised and expanded edition

The Hidden Forces That Shape Our Decisions

Dan Ariely

HARPER

An Imprint of HarperCollins*Publishers*

www.harpercollins.com

HarperCollins books may be purchased for educational, business, or
sales promotional use. For information, please write: Special Markets
Department, HarperCollins Publishers, 10 East 53rd Street, New York,
NY 10022.

Originally published in 2008, in a different format, by HarperCollins
Publishers.

FIRST REVISED AND EXPANDED EDITION

Library of Congress Cataloging-in-Publication Data is available upon
request.

ISBN: 978-0-06-185454-5

09 10 11 12 13 ID/RRD 10 9 8 7 6 5 4 3 2 1

To my mentors, colleagues, and students—
who make research exciting

Contents

contents

A Note to Readers

*Dear readers, friends, and
social science enthusiasts,*

Welcome to the revised and expanded edition of *Predictably Irrational.*

Since my early days as a patient in the burn department,* I have been acutely aware that humans engage in actions and make decisions that are often divorced from rationality, and sometimes very far from ideal. Over the years I've tried to understand the silly, dumb, odd, amusing, and sometimes dangerous mistakes we all make, in the hope that by understanding our irrational quirks, we can retrain ourselves to make better decisions.

My theoretical and applied interest in irrationality has guided me to the emerging field of behavioral economics, where I've embraced these quirks as a fundamental element of human behavior. In my research, I've looked at a range of human foibles, asking questions such as these: Why do we

---

*For more on what happened, see the Introduction.

get overexcited when something is FREE!? What role do emotions play in our decisions? How does procrastination play games with us? What are the functions of our strange social norms? Why do we hang on to false beliefs despite evidence to the contrary? Trying to answer these questions has provided me with endless hours of fun, and the new understanding that it brought has changed my professional and personal life.

The experiments my colleagues and I have conducted helped us discover why our participants (and humans in general, including ourselves) fail to reason properly. It's been satisfying to try to understand why we act the way we do, and fun to share our findings with people who have also wondered about their own decisions.

NEVERTHELESS, BEFORE THE financial crisis of 2008, I'd hit a lot of roadblocks when trying to expand on the implications of our ideas, experiments, and findings. For example, after I gave a presentation at a conference, a fellow I'll call Mr. Logic (a composite of many people I have debated with over the years) buttonholed me.

"I enjoy hearing about all the different kinds of small-scale irrationalities that you demonstrate in your experiments," he told me, handing me his card. "They're quite interesting—great stories for cocktail parties." He paused. "But you don't understand how things work in the real world. Clearly, when it comes to making important decisions, all of these irrationalities disappear, because when it truly matters, people think carefully about their options before they act. And certainly when it comes to the stock market, where the decisions are critically important, all these irrationalities go away and rationality prevails."

This type of sentiment has not been restricted to Chicago economists—the elite of rational economic thought. I have often been amazed at the prevalence of this sentiment (I'd even dare to call it indoctrination) among people who have no particular training in economics. Somehow, the basic ideas of economics and the belief in overarching rationality have become so ingrained in our understanding of the social world around us that people from all walks of life seemed to accept them as basic laws of nature. When it came to the stock market, rationality and economics were thought to be as perfect a match as Fred Astaire and Ginger Rogers.

Whenever I've been confronted with this type of criticism, I would try to dig a bit deeper and inquire why the belief in rationality surfaced whenever people made decisions in the stock market. My conversational partner would usually try patiently to persuade me to his way of thinking. "Don't you understand," Mr. Logic would say, "that when there is a lot of money on the table, people think especially hard about their options and do their best to maximize their returns."

"Doing their best," I would say in rebuttal, "is not the same as being able to make optimal decisions. What about individual investors, who put all their money in their own company's stock, don't diversify enough,* and lose a substantial part of their fortune? What about people who are approaching their sixtieth birthday and still don't contribute to their 401(k)s? They're giving up free money, because they can withdraw it, along with the match from their company, almost immediately.[1]

---

* One of the main lessons of finance is that diversification is very important. When we work for a company, we already have a lot invested in it, in terms of our salary, so investing even more in the same company is very bad in terms of diversification.

"OK," he would reluctantly agree. "It's true that sometimes individual investors make mistakes, but professional investors must, by definition, act rationally because they deal with a lot of money and are paid to maximize their returns. On top of that, they work in a competitive environment that keeps them on their toes and ensures that they will always make normatively correct decisions."

"Do you really want to argue," I would ask, squinting at him, "that just because they are acting in their own best interests, professional investors never make big mistakes?"

"Not all the time," Mr. Logic would calmly reply, "but in the aggregate they make normatively correct decisions. One person makes a random mistake in this direction, another makes a mistake in the other direction, and, collectively, all these mistakes cancel each other out—keeping the pricing in the market optimal."

At this point in the conversation, I must admit, my patience would start to wear out. "What makes you think," I would ask, "that the mistakes people make—even if those people are professional investors—are simply random? Think about Enron. Enron's auditors were involved in substantial conflicts of interest, which ultimately led them to turn a blind eye (or perhaps two blind eyes, a stuffed nose, and plugged ears) to what was happening inside the company. Or what about the incentives of money managers, who make big bucks when their clients do, but don't lose anything when the opposite happens? In such environments, where misaligned incentives and conflicts of interest are endemic, people would most likely make the same mistakes over and over, and these mistakes would not cancel each other out. In fact, these mistakes are the most dangerous because they aren't random at all, and in the aggregate, can be devastating to the economy."

At this point, Mr. Logic would take out the final weapon from his rational arsenal and remind me about *(Zap! Pow!)* the force of arbitrage—the magical power that eliminates the effects of individuals' mistakes and makes the market, as a whole, act perfectly rationally. How does arbitrage fix the market? When the markets are free and frictionless—and even if most investors are irrational—a small set of supersmart, rational investors will take advantage of everyone else's poor decisions (for example, they might buy a stock from those of us who mistakenly undervalue it), and in the process of competing for a bigger piece of the pie make a lot of money for themselves and restore market pricing to its rational and correct levels. "Arbitrage is the reason why your notion of behavioral economics is wrong," Mr. Logic would tell me triumphantly.

Sadly, arbitrage is not an idea we can test empirically, because we cannot run one version of the stock market consisting of Joe Schmoes like you and me and another one consisting of Joe Schmoes plus some of these extra-special, super-rational investors—these Supermen who save the financial world from danger every day, while retaining their anonymous Clark Kent identities.

I wish I could tell you that I would often persuade my conversational partner to accept my point of view, but in almost all cases it would become very clear that neither of us was going to be converted to the other's viewpoint. Of course, I ran into the biggest difficulties when arguing for irrationality with card-carrying rational economists, whose disregard of my experimental data was almost as intense as their nearly religious belief in rationality (if Adam Smith's "invisible hand" doesn't sound like God, I don't know what does). This basic sentiment was expressed succinctly by two fabulous

Chicago economists, Steven Levitt and John List, suggesting that the practical usefulness of behavioral economics has been shown to be marginal at best:

> *Perhaps the greatest challenge facing behavioral economics is demonstrating its applicability in the real world. In nearly every instance, the strongest empirical evidence in favor of behavioral anomalies emerges from the lab. Yet there are many reasons to suspect that these laboratory findings might fail to generalize to real markets. . . . For example, the competitive nature of markets encourages individualistic behavior and selects for participants with those tendencies. Compared to lab behavior, therefore, the combination of market forces and experience might lessen the importance of these qualities in everyday markets.*[2]

Given these kinds of responses, I am often left scratching my head, wondering why so many smart people are convinced that irrationality disappears when it comes to important decisions about money. Why do they assume that institutions, competition, and market mechanisms can inoculate us against mistakes? If competition was sufficient to overcome irrationality, wouldn't that eliminate brawls in sporting competitions, or the irrational self-destructive behaviors of professional athletes? What is it about circumstances involving money and competition that might make people more rational? Do the defenders of rationality believe that we have different brain mechanisms for making small versus large decisions and yet another yet another for dealing with the stock market? Or do they simply have a bone-deep belief that the invisible hand and the wisdom of the markets guarantee optimal behavior under all conditions?

As a social scientist, I'm not sure which model describing human behavior in markets—rational economics, behavioral economics, or something else—is best, and I wish we could set up a series of experiments to figure this out. Unfortunately, since it is basically impossible to do any real experiments with the stock market, I've been left befuddled by the deep conviction in the rationality of the market. And I've wondered if we really want to build our financial institutions, our legal system, and our policies on such a foundation.

As I WAS asking myself these questions, something very big happened.

Soon after *Predictably Irrational* was published, in early 2008, the financial world blew to smithereens, like something in a science fiction movie.* Alan Greenspan, the formerly much-worshipped chairman of the Federal Reserve, told Congress in October 2008 that he was "shocked" (shocked!) that the markets did not work as anticipated, or automatically self-correct as they were supposed to. He said he made a mistake in assuming that the self-interest of organizations, specifically banks and others, was such that they were capable of protecting their own shareholders.

For my part, I was shocked that Greenspan, one of the tireless advocates of deregulation and a true believer in letting market forces have their way, would publicly admit that

---

*I don't think that there was any causal connection between the publication of *Predictably Irrational* and the financial meltdown, but you must admit that the timing is curious.

his assumptions about the rationality of markets were wrong. A few months before this confession, I could never have imagined that Greenspan would utter such a statement. Aside from feeling vindicated, I also felt that Greenspan's confession was an important step forward. After all, they say that the first step toward recovery is admitting you have a problem.

Still, the terrible loss of homes and jobs has been a very high price to pay for learning that we might not be as rational as Greenspan and other traditional economists had thought. What we've learned is that relying on standard economic theory alone as a guiding principle for building markets and institutions might, in fact, be dangerous. It has become tragically clear that the mistakes we all make are not at all random, but part and parcel of the human condition. Worse, our mistakes of judgment can aggregate in the market, sparking a scenario in which, much like an earthquake, no one has any idea what is happening. (Al Roth, an economist at Harvard, and one of the smartest people I know, has summarized this issue by saying, "In theory, there is no difference between theory and practice, but in practice there is a great deal of difference.")

A few days after Greenspan's congressional testimony, the *New York Times* columnist David Brooks wrote that Greenspan's confession would " . . . amount to a coming-out party for behavioral economists and others who are bringing sophisticated psychology to the realm of public policy. At least these folks have plausible explanations for why so many people could have been so gigantically wrong about the risks they were taking."[3]

All of a sudden, it looked as if some people were beginning to understand that the study of small-scale mistakes was not just a source for amusing dinner-table anecdotes. I felt both exonerated and relieved.

While this is a very depressing time for the economy as a whole, and for all of us individually, the turnabout on Greenspan's part has created new opportunities for behavioral economics, and for those willing to learn and alter the way they think and behave. From crisis comes opportunity, and perhaps this tragedy will cause us to finally accommodate new ideas, and—I hope—begin to rebuild.

WRITING A BOOK in the age of blogging and e-mail is an absolute treat because I get continuous feedback from readers, which causes me to learn about, reconsider, and rethink different aspects of human behavior. I've also had some very interesting discussions with readers about the links between behavioral economics and what's happening in the financial markets, and about random topics relating to everyday irrationalities.

At the end of this book (following the material originally included in *Predictably Irrational*), I offer a few reflections and anecdotes about some of the chapters in the book, and my thoughts about the financial markets—what got us into this mess, how we can understand it from the perspective of behavioral economics, and how we can try to get out of it.

First, however, let's explore some of our irrationalities.

Introduction

*How an Injury Led Me to Irrationality and
to the Research Described Here*

I have been told by many people that I have an unusual way
of looking at the world. Over the last 20 years or so of my
research career, it's enabled me to have a lot of fun figuring out
what really influences our decisions in daily life (as opposed to
what we think, often with great confidence, influences them).

Do you know why we so often promise ourselves to diet,
only to have the thought vanish when the dessert cart rolls
by?

Do you know why we sometimes find ourselves excitedly
buying things we don't really need?

Do you know why we still have a headache after taking a
one-cent aspirin, but why that same headache vanishes when
the aspirin costs 50 cents?

Do you know why people who have been asked to recall
the Ten Commandments tend to be more honest (at least immediately afterward) than those who haven't? Or why honor
codes actually do reduce dishonesty in the workplace?

By the end of this book, you'll know the answers to these and many other questions that have implications for your personal life, for your business life, and for the way you look at the world. Understanding the answer to the question about aspirin, for example, has implications not only for your choice of drugs, but for one of the biggest issues facing our society: the cost and effectiveness of health insurance. Understanding the impact of the Ten Commandments in curbing dishonesty might help prevent the next Enron-like fraud. And understanding the dynamics of impulsive eating has implications for every other impulsive decision in our lives—including why it's so hard to save money for a rainy day.

My goal, by the end of this book, is to help you fundamentally rethink what makes you and the people around you tick. I hope to lead you there by presenting a wide range of scientific experiments, findings, and anecdotes that are in many cases quite amusing. Once you see how systematic certain mistakes are—how we repeat them again and again—I think you will begin to learn how to avoid some of them.

But before I tell you about my curious, practical, entertaining (and in some cases even delicious) research on eating, shopping, love, money, procrastination, beer, honesty, and other areas of life, I feel it is important that I tell you about the origins of my somewhat unorthodox worldview—and therefore of this book. Tragically, my introduction to this arena started with an accident many years ago that was anything but amusing.

ON WHAT WOULD otherwise have been a normal Friday afternoon in the life of an eighteen-year-old Israeli, everything changed irreversibly in a matter of a few seconds. An explosion of a large magnesium flare, the kind used to illuminate battlefields at night, left 70 percent of my body covered with third-degree burns.

The next three years found me wrapped in bandages in a hospital and then emerging into public only occasionally, dressed in a tight synthetic suit and mask that made me look like a crooked version of Spider-Man. Without the ability to participate in the same daily activities as my friends and family, I felt partially separated from society and as a consequence started to observe the very activities that were once my daily routine as if I were an outsider. As if I had come from a different culture (or planet), I started reflecting on the goals of different behaviors, mine and those of others. For example, I started wondering why I loved one girl but not another, why my daily routine was designed to be comfortable for the physicians but not for me, why I loved going rock climbing but not studying history, why I cared so much about what other people thought of me, and mostly what it is about life that motivates people and causes us to behave as we do.

During the years in the hospital following my accident, I had extensive experience with different types of pain and a great deal of time between treatments and operations to reflect on it. Initially, my daily agony was largely played out in the "bath," a procedure in which I was soaked in disinfectant solution, the bandages were removed, and the dead particles of skin were scraped off. When the skin is intact, disinfectants create a low-level sting, and in general the bandages come off easily. But when there is little or no skin—as in my case because of my extensive burns—the disinfectant stings unbearably, the bandages stick to the flesh, and removing them (often tearing them) hurts like nothing else I can describe.

Early on in the burn department I started talking to the

nurses who administered my daily bath, in order to understand their approach to my treatment. The nurses would
routinely grab hold of a bandage and rip it off as fast as possible, creating a relatively short burst of pain; they would repeat this process for an hour or so until they had removed
every one of the bandages. Once this process was over I was
covered with ointment and with new bandages, in order to
repeat the process again the next day.

The nurses, I quickly learned, had theorized that a vigor-
ous tug at the bandages, which caused a sharp spike of pain,
was preferable (to the patient) to a slow pulling of the wrap-
pings, which might not lead to such a severe spike of pain but
would extend the treatment, and therefore be more painful
overall. The nurses had also concluded that there was no difference between two possible methods: starting at the most
painful part of the body and working their way to the least
painful part; or starting at the least painful part and advanc-
ing to the most excruciating areas.

As someone who had actually experienced the pain of the
bandage removal process, I did not share their beliefs (which
had never been scientifically tested). Moreover, their theories
gave no consideration to the amount of fear that the patient
felt anticipating the treatment; to the difficulties of dealing
with fluctuations of pain over time; to the unpredictability of
not knowing when the pain will start and ease off; or to the
benefits of being comforted with the possibility that the pain
would be reduced over time. But, given my helpless position,
I had little influence over the way I was treated.

As soon as I was able to leave the hospital for a prolonged
period (I would still return for occasional operations and
treatments for another five years), I began studying at Tel
Aviv University. During my first semester, I took a class that

profoundly changed my outlook on research and largely determined my future. This was a class on the physiology of the brain, taught by professor Hanan Frenk. In addition to the fascinating material Professor Frenk presented about the workings of the brain, what struck me most about this class was his attitude to questions and alternative theories. Many times, when I raised my hand in class or stopped by his office to suggest a different interpretation of some results he had presented, he replied that my theory was indeed a possibility (somewhat unlikely, but a possibility nevertheless)—and would then challenge me to propose an empirical test to distinguish it from the conventional theory.

Coming up with such tests was not easy, but the idea that science is an empirical endeavor in which all the participants, including a new student like myself, could come up with alternative theories, as long as they found empirical ways to test these theories, opened up a new world to me. On one of my visits to Professor Frenk's office, I proposed a theory explaining how a certain stage of epilepsy developed, and included an idea for how one might test it in rats.

Professor Frenk liked the idea, and for the next three months I operated on about 50 rats, implanting catheters in their spinal cords and giving them different substances to create and reduce their epileptic seizures. One of the practical problems with this approach was that the movements of my hands were very limited, because of my injury, and as a consequence it was very difficult for me to operate on the rats. Luckily for me, my best friend, Ron Weisberg (an avid vegetarian and animal lover), agreed to come with me to the lab for several weekends and help me with the procedures—a true test of friendship if ever there was one.

In the end, it turned out that my theory was wrong, but

this did not diminish my enthusiasm. I was able to learn something about my theory, after all, and even though the theory was wrong, it was good to know this with high certainty. I always had many questions about how things work and how people behave, and my new understanding—that science provides the tools and opportunities to examine anything I found interesting—lured me into the study of how people behave.

With these new tools, I focused much of my initial efforts on understanding how we experience pain. For obvious reasons I was most concerned with such situations as the bath treatment, in which pain must be delivered to a patient over a long period of time. Was it possible to reduce the overall agony of such pain? Over the next few years I was able to carry out a set of laboratory experiments on myself, my friends, and volunteers—using physical pain induced by heat, cold water, pressure, loud sounds, and even the psychological pain of losing money in the stock market—to probe for the answers.

By the time I had finished, I realized that the nurses in the burn unit were kind and generous individuals (well, there was one exception) with a lot of experience in soaking and removing bandages, but they still didn't have the right theory about what would minimize their patients' pain. How could they be so wrong, I wondered, considering their vast experience? Since I knew these nurses personally, I knew that their behavior was not due to maliciousness, stupidity, or neglect. Rather, they were most likely the victims of inherent biases in their perceptions of their patients' pain—biases that apparently were not altered even by their vast experience.

For these reasons, I was particularly excited when I returned to the burn department one morning and presented my results, in the hope of influencing the bandage removal procedures for other patients. It turns out, I told the nurses and physicians, that people feel less pain if treatments (such as removing bandages in a bath) are carried out with lower intensity and longer duration than if the same goal is achieved through high intensity and a shorter duration. In other words, I would have suffered less if they had pulled the bandages off slowly rather than with their quick-pull method.

The nurses were genuinely surprised by my conclusions, but I was equally surprised by what Etty, my favorite nurse, had to say. She admitted that their understanding had been lacking and that they should change their methods. But she also pointed out that a discussion of the pain inflicted in the bath treatment should also take into account the psychological pain that the nurses experienced when their patients screamed in agony. Pulling the bandages quickly might be more understandable, she explained, if it were indeed the nurses' way of shortening their own torment (and their faces often did reveal that they were suffering). In the end, though, we all agreed that the procedures should be changed, and indeed, some of the nurses followed my recommendations.

My recommendations never changed the bandage removal process on a greater scale (as far as I know), but the episode left a special impression on me. If the nurses, with all their experience, misunderstood what constituted reality for the patients they cared so much about, perhaps other people similarly misunderstand the consequences of their behaviors and, for that reason, repeatedly make the wrong decisions. I decided to expand my scope of research, from pain to the examination of cases in which individuals make repeated mistakes—without being able to learn much from their experiences.

THIS JOURNEY INTO the many ways in which we are all ir-rational, then, is what this book is about. The discipline that allows me to play with this subject matter is called *behavioral economics*, or judgment and decision making (JDM).

Behavioral economics is a relatively new field, one that draws on aspects of both psychology and economics. It has led me to study everything from our reluctance to save for retirement to our inability to think clearly during sexual arousal. It's not just the behavior that I have tried to under-stand, though, but also the decision-making processes behind such behavior—yours, mine, and everybody else's. Before I go on, let me try to explain, briefly, what behavioral eco-nomics is all about and how it is different from standard economics. Let me start out with a bit of Shakespeare:

> *What a piece of work is a man! how noble in reason!*
> *how infinite in faculty! in form and moving how*
> *express and admirable! in action how like an angel!*
> *in apprehension how like a god! The beauty of the*
> *world, the paragon of animals.* —from Act II,
> scene 2, of *Hamlet*

The predominant view of human nature, largely shared by economists, policy makers, nonprofessionals, and every-day Joes, is the one reflected in this quotation. Of course, this view is largely correct. Our minds and bodies are capable of amazing acts. We can see a ball thrown from a distance, instantly calculate its trajectory and impact, and then move our body and hands in order to catch it. We can learn new languages with ease, particularly as young children. We can master chess. We can recognize thousands of faces without confusing them. We can produce music, literature, technology, and art—and the list goes on and on.

Shakespeare is not alone in his appreciation for the human mind. In fact, we all think of ourselves along the lines of Shakespeare's depiction (although we do realize that our neighbors, spouses, and bosses do not always live up to this standard). Within the domain of science, these assumptions about our ability for perfect reasoning have found their way into economics. In economics, this very basic idea, called *rationality*, provides the foundation for economic theories, predictions, and recommendations.

From this perspective, and to the extent that we all believe in human rationality, we are all economists. I don't mean that each of us can intuitively develop complex game-theoretical models or understand the generalized axiom of revealed preference (GARP); rather, I mean that we hold the basic beliefs about human nature on which economics is built. In this book, when I mention the *rational* economic model, I refer to the basic assumption that most economists and many of us hold about human nature—the simple and compelling idea that we are capable of making the right decisions for ourselves.

Although a feeling of awe at the capability of humans is clearly justified, there is a large difference between a deep sense of admiration and the assumption that our reasoning abilities are perfect. In fact, this book is about human *irrationality*—about our distance from perfection. I believe that recognizing where we depart from the ideal is an important part of the quest to truly understand ourselves, and one that promises many practical benefits. Understanding irrationality is important for our everyday actions and decisions, and for understanding how we design our environment and the choices it presents to us.

My further observation is that we are not only irrational, but *predictably irrational*—that our irrationality happens the same way, again and again. Whether we are acting as consumers, businesspeople, or policy makers, understanding how we are predictably irrational provides a starting point for improving our decision making and changing the way we live for the better.

This leads me to the real "rub" (as Shakespeare might have called it) between conventional economics and behavioral economics. In conventional economics, the assumption that we are all rational implies that, in everyday life, we compute the value of all the options we face and then follow the best possible path of action. What if we make a mistake and do something irrational? Here, too, traditional economics has an answer: "market forces" will sweep down on us and swiftly set us back on the path of righteousness and rationality. On the basis of these assumptions, in fact, generations of economists since Adam Smith have been able to develop far-reaching conclusions about everything from taxation and health-care policies to the pricing of goods and services.

But, as you will see in this book, we are really far less rational than standard economic theory assumes. Moreover, these irrational behaviors of ours are neither random nor senseless. They are systematic, and since we repeat them again and again, predictable. So, wouldn't it make sense to modify standard economics, to move it away from naive psychology (which often fails the tests of reason, introspection, and—most important—empirical scrutiny)? This is exactly what the emerging field of behavioral economics, and this book as a small part of that enterprise, is trying to accomplish.

As YOU WILL see in the pages ahead, each of the chapters in this book is based on a few experiments I carried out over the years with some terrific colleagues (at the end of the book, I have included short biographies of my amazing collaborators). Why experiments? Life is complex, with multiple forces simultaneously exerting their influences on us, and this complexity makes it difficult to figure out exactly how each of these forces shapes our behavior. For social scientists, experiments are like microscopes or strobe lights. They help us slow human behavior to a frame-by-frame narration of events, isolate individual forces, and examine those forces carefully and in more detail. They let us test directly and unambiguously what makes us tick.

There is one other point I want to emphasize about experiments. If the lessons learned in any experiment were limited to the exact environment of the experiment, their value would be limited. Instead, I would like you to think about experiments as an illustration of a general principle, providing insight into how we think and how we make decisions—not only in the context of a particular experiment but, by extrapolation, in many contexts of life.

In each chapter, then, I have taken a step in extrapolating the findings from the experiments to other contexts, attempting to describe some of their possible implications for life, business, and public policy. The implications I have drawn are, of course, just a partial list.

To get real value from this, and from social science in general, it is important that you, the reader, spend some time thinking about how the principles of human behavior identified in the experiments apply to your life. My suggestion to you is to pause at the end of each chapter and consider

introduction

whether the principles revealed in the experiments might make your life better or worse, and more importantly what you could do differently, given your new understanding of human nature. This is where the real adventure lies.

And now for the journey.

xxxii

CHAPTER 1

The Truth about Relativity

*Why Everything Is Relative—Even
When It Shouldn't Be*

One day while browsing the World Wide Web (obviously for work—not just wasting time), I stumbled on the following ad, on the Web site of a magazine, the *Economist*.

Economist.com	SUBSCRIPTIONS

OPINION
WORLD
BUSINESS
FINANCE & ECONOMICS
SCIENCE & TECHNOLOGY
PEOPLE
BOOKS & ARTS
MARKETS & DATA
DIVERSIONS
	Welcome to
	The Economist Subscription Centre
	
	Pick the type of subscription you want to buy or renew.
	
	❑ **Economist.com subscription** - US $59.00
	One-year subscription to Economist.com.
	  Includes online access to all articles from
	  The Economist since 1997.
	
	❑ **Print subscription** - US $125.00
	One-year subscription to the print edition
	of *The Economist*.
	
	❑ **Print & web subscription** - US $125.00
	One-year subscription to the print edition
	of *The Economist* and online access to all
	  articles from *The Economist* since 1997.

I read these offers one at a time. The first offer—the Internet subscription for $59—seemed reasonable. The second option—the $125 print subscription—seemed a bit expensive, but still reasonable.

But then I read the third option: a print *and* Internet subscription for $125. I read it twice before my eye ran back to the previous options. Who would want to buy the print option alone, I wondered, when both the Internet and the print subscriptions were offered for the same price? Now, the print-only option may have been a typographical error, but I suspect that the clever people at the *Economist*'s London offices (and they are clever—and quite mischievous in a British sort of way) were actually manipulating me. I am pretty certain that they wanted me to skip the Internet-only option (which they assumed would be my choice, since I was reading the advertisement on the Web) and jump to the more expensive option: Internet and print.

But how could they manipulate me? I suspect it's because the *Economist*'s marketing wizards (and I could just picture them in their school ties and blazers) knew something important about human behavior: humans rarely choose things in absolute terms. We don't have an internal value meter that tells us how much things are worth. Rather, we focus on the relative advantage of one thing over another, and estimate value accordingly. (For instance, we don't know how much a six-cylinder car is worth, but we can assume it's more expensive than the four-cylinder model.)

In the case of the *Economist*, I may not have known whether the Internet-only subscription at $59 was a better deal than the print-only option at $125. But I certainly knew that the print-and-Internet option for $125 was better than the print-only option at $125. In fact, you could reasonably deduce that in the combination package, the Internet subscription is free! "It's a bloody steal—go for it, governor!" I could almost hear them shout from the riverbanks of the Thames. And I have to admit, if I had been inclined to subscribe I probably would have taken the package deal myself. (Later, when I tested the offer on a large number of participants, the vast majority preferred the Internet-and-print deal.)

So what was going on here? Let me start with a fundamental observation: most people don't know what they want unless they see it in context. We don't know what kind of racing bike we want—until we see a champ in the Tour de France ratcheting the gears on a particular model. We don't know what kind of speaker system we like—until we hear a set of speakers that sounds better than the previous one. We don't even know what we want to do with our lives—until we find a relative or a friend who is doing just what we think we should be doing. Everything is relative, and that's the point. Like an airplane pilot landing in the dark, we want runway lights on either side of us, guiding us to the place where we can touch down our wheels.

In the case of the *Economist*, the decision between the Internet-only and print-only options would take a bit of thinking. Thinking is difficult and sometimes unpleasant. So the *Economist*'s marketers offered us a no-brainer: relative to the print-only option, the print-and-Internet option looks clearly superior.

The geniuses at the *Economist* aren't the only ones who understand the importance of relativity. Take Sam, the television salesman. He plays the same general type of trick on us when he decides which televisions to put together on display:

36-inch Panasonic for $690
42-inch Toshiba for $850
50-inch Philips for $1,480

Which one would you choose? In this case, Sam knows that customers find it difficult to compute the value of different options. (Who really knows if the Panasonic at $690 is a better deal than the Philips at $1,480?) But Sam also knows that given three choices, most people will take the middle choice (as in landing your plane between the runway lights). So guess which television Sam prices as the middle option? That's right—the one he wants to sell!

Of course, Sam is not alone in his cleverness. The *New York Times* ran a story recently about Gregg Rapp, a restaurant consultant, who gets paid to work out the pricing for menus. He knows, for instance, how lamb sold this year as opposed to last year; whether lamb did better paired with squash or with risotto; and whether orders decreased when the price of the main course was hiked from $39 to $41.

One thing Rapp has learned is that high-priced entrées on the menu boost revenue for the restaurant—even if no one buys them. Why? Because even though people generally won't buy the most expensive dish on the menu, they will order the second most expensive dish. Thus, by creating an expensive dish, a restaurateur can lure customers into ordering the second most expensive choice (which can be cleverly engineered to deliver a higher profit margin).[4]

So LET'S RUN through the *Economist*'s sleight of hand in slow motion.

As you recall, the choices were:

1. Internet-only subscription for $59.
2. Print-only subscription for $125.
3. Print-and-Internet subscription for $125.

When I gave these options to 100 students at MIT's Sloan School of Management, they opted as follows:

1. Internet-only subscription for $59—**16 students**
2. Print-only subscription for $125—**zero students**
3. Print-and-Internet subscription for $125—**84 students**

So far these Sloan MBAs are smart cookies. They all saw the advantage in the print-and-Internet offer over the print-only offer. But were they influenced by the mere presence of the print-only option (which I will henceforth, and for good reason, call the "decoy"). In other words, suppose that I removed the decoy so that the choices would be the ones seen in the figure below:

Economist.com	SUBSCRIPTIONS
OPINION	**Welcome to**
WORLD	**The Economist Subscription Centre**
BUSINESS	
FINANCE & ECONOMICS	Pick the type of subscription you want to buy or renew.
SCIENCE & TECHNOLOGY	
PEOPLE	❑ **Economist.com subscription** - US $59.00
BOOKS & ARTS	One-year subscription to Economist.com.
MARKETS & DATA	Includes online access to all articles from
DIVERSIONS	*The Economist* since 1997.
	❑ **Print & web subscription** - US $125.00 One-year subscription to the print edition of *The Economist* and online access to all articles from *The Economist* since 1997.

Would the students respond as before (16 for the Internet only and 84 for the combination)?

Certainly they would react the same way, wouldn't they? After all, the option I took out was one that no one selected, so it should make no difference. Right?

Au contraire! This time, 68 of the students chose the Internet-only option for $59, up from 16 before. And only 32 chose the combination subscription for $125, down from 84 before.*

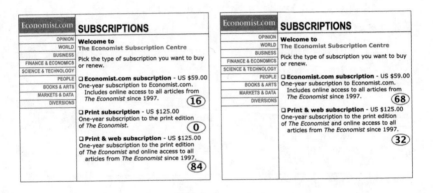

What could have possibly changed their minds? Nothing rational, I assure you. It was the mere presence of the decoy that sent 84 of them to the print-and-Internet option (and 16 to the Internet-only option). And the absence of the decoy had them choosing differently, with 32 for print-and-Internet and 68 for Internet-only.

This is not only irrational but predictably irrational as well. Why? I'm glad you asked.

---

*As a convention in this book, every time I mention that conditions are different from each other, it is always a statistically significant difference. I refer the interested reader to the end of this book for a list of the original academic papers and additional readings.

LET ME OFFER you this visual demonstration of relativity.

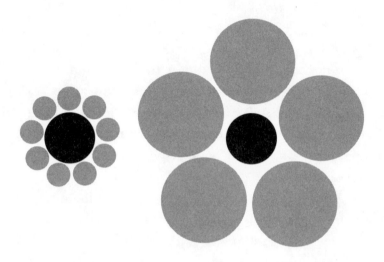

As you can see, the middle circle can't seem to stay the same size. When placed among the larger circles, it gets smaller. When placed among the smaller circles, it grows bigger. The middle circle is the same size in both positions, of course, but it appears to change depending on what we place next to it.

This might be a mere curiosity, but for the fact that it mirrors the way the mind is wired: we are always looking at the things around us in relation to others. We can't help it. This holds true not only for physical things—toasters, bicycles, puppies, restaurant entrées, and spouses—but for experiences such as vacations and educational options, and for ephemeral things as well: emotions, attitudes, and points of view.

We always compare jobs with jobs, vacations with vacations, lovers with lovers, and wines with wines. All this relativity reminds me of a line from the film *Crocodile Dundee,* when a street hoodlum pulls a switchblade against our hero, Paul Hogan. "You call that a knife?" says Hogan incredulously, withdrawing a bowie blade from the back of his boot. "Now *this*," he says with a sly grin, "is a knife."

RELATIVITY IS (RELATIVELY) easy to understand. But there's one aspect of relativity that consistently trips us up. It's this: we not only tend to compare things with one another but also tend to focus on comparing things that are easily comparable—and avoid comparing things that cannot be compared easily.

That may be a confusing thought, so let me give you an example. Suppose you're shopping for a house in a new town. Your real estate agent guides you to three houses, all of which interest you. One of them is a contemporary, and two are colonials. All three cost about the same; they are all equally desirable; and the only difference is that one of the colonials (the "decoy") needs a new roof and the owner has knocked a few thousand dollars off the price to cover the additional expense.

So which one will you choose?

The chances are good that you will *not* choose the contemporary and you will *not* choose the colonial that needs the new roof, but you will choose the other colonial. Why? Here's the rationale (which is actually quite irrational). We like to make decisions based on comparisons. In the case of the three houses, we don't know much about the contemporary (we don't have another house to compare it with), so that house goes on the sidelines. But we do know that one of the colonials is better than the other one. That is, the colonial with the good roof is better than the one with the bad roof. Therefore, we will reason that it is better overall and go for the colonial with the good roof, spurning the contemporary and the colonial that needs a new roof.

To better understand how relativity works, consider the following illustration:

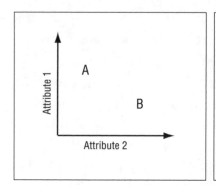 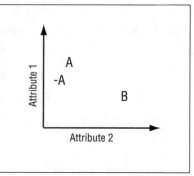

In the left side of this illustration we see two options, each of which is better on a different attribute. Option (A) is better on attribute 1—let's say quality. Option (B) is better on attribute 2—let's say beauty. Obviously these are two very different options and the choice between them is not simple. Now consider what happens if we add another option, called (−A) (see the right side of the illustration). This option is clearly worse than option (A), but it is also very similar to it, making the comparison between them easy, and suggesting that (A) is not only better than (−A) but also better than (B).

In essence, introducing (−A), the decoy, creates a simple relative comparison with (A), and hence makes (A) look better, not just relative to (−A), but overall as well. As a consequence, the inclusion of (−A) in the set, even if no one ever selects it, makes people more likely to make (A) their final choice.

Does this selection process sound familiar? Remember the pitch put together by the *Economist*? The marketers there knew that we didn't know whether we wanted an Internet subscription or a print subscription. But they figured that, of the three options, the print-and-Internet combination would be the offer we would take.

Here's another example of the decoy effect. Suppose you are planning a honeymoon in Europe. You've already decided to go to one of the major romantic cities and have narrowed your choices to Rome and Paris, your two favorites. The travel agent presents you with the vacation packages for each city, which includes airfare, hotel accommodations, sightseeing tours, and a free breakfast every morning. Which would you select?

For most people, the decision between a week in Rome and a week in Paris is not effortless. Rome has the Coliseum; Paris, the Louvre. Both have a romantic ambience, fabulous food, and fashionable shopping. It's not an easy call. But suppose you were offered a third option: Rome without the free breakfast, called –Rome or the decoy.

If you were to consider these three options (Paris, Rome, –Rome), you would immediately recognize that whereas Rome with the free breakfast is about as appealing as Paris with the free breakfast, the inferior option, which is Rome without the free breakfast, is a step down. The comparison between the clearly inferior option (–Rome) makes Rome with the free breakfast seem even better. In fact, –Rome makes Rome with the free breakfast look so good that you judge it to be even better than the difficult-to-compare option, Paris with the free breakfast.

ONCE YOU SEE the decoy effect in action, you realize that it is the secret agent in more decisions than we could imagine. It even helps us decide whom to date—and, ultimately, whom to marry. Let me describe an experiment that explored just this subject.

As students hurried around MIT one cold weekday, I asked some of them whether they would allow me to take their pictures for a study. In some cases, I got disapproving looks. A few students walked away. But most of them were happy to participate, and before long, the card in my digital camera was filled with images of smiling students. I returned to my office and printed 60 of them—30 of women and 30 of men.

The following week I made an unusual request of 25 of my undergraduates. I asked them to pair the 30 photographs of men and the 30 of women by physical attractiveness (matching the men with other men, and the women with other women). That is, I had them pair the Brad Pitts and the George Clooneys of MIT, as well as the Woody Allens and the Danny DeVitos (sorry, Woody and Danny). Out of these 30 pairs, I selected the six pairs—three female pairs and three male pairs—that my students seemed to agree were most alike.

Now, like Dr. Frankenstein himself, I set about giving these faces my special treatment. Using Photoshop, I mutated the pictures just a bit, creating a slightly but noticeably less attractive version of each of them. I found that just the slightest movement of the nose threw off the symmetry. Using another tool, I enlarged one eye, eliminated some of the hair, and added traces of acne.

No flashes of lightning illuminated my laboratory; nor was there a baying of the hounds on the moor. But this was still a good day for science. By the time I was through, I had the MIT equivalent of George Clooney in his prime (A) and the MIT equivalent of Brad Pitt in his prime (B), and also a George Clooney with a slightly drooping eye and thicker nose (−A, the decoy) and a less symmetrical version of Brad Pitt (−B, another decoy). I followed the same procedure for the less attractive pairs. I had the MIT equivalent of Woody

Allen with his usual lopsided grin (A) and Woody Allen with an unnervingly misplaced eye (–A), as well as Danny DeVito (B) and a slightly disfigured version of Danny DeVito (–B).

For each of the 12 photographs, in fact, I now had a regular version as well as an inferior (–) decoy version. (See the illustration for an example of the two conditions used in the study.)

It was now time for the main part of the experiment. I took all the sets of pictures and made my way over to the student union. Approaching one student after another, I asked each to participate. When the students agreed, I handed them a sheet with three pictures (as in the illustration here). Some of them had the regular picture (A), the decoy of that picture (–A), and the other regular picture (B). Others had the regular picture (B), the decoy of that picture (–B), and the other regular picture (A).

For example, a set might include a regular Clooney (A), a decoy Clooney (–A), and a regular Pitt (B); or a regular Pitt (B), a decoy Pitt (–B), and a regular Clooney (A). After selecting a sheet with either male or female pictures, according to their preferences, I asked the students to circle the people they would pick to go on a date with, if they had a choice. All this took quite a while, and when I was done, I had distributed 600 sheets.

What was my motive in all this? Simply to determine if the existence of the distorted picture (–A or –B) would push my participants to choose the similar but undistorted picture. In other words, would a slightly less attractive George Clooney (–A) push the participants to choose the perfect George Clooney over the perfect Brad Pitt?

There were no pictures of Brad Pitt or George Clooney in my experiment, of course. Pictures (A) and (B) showed ordi-

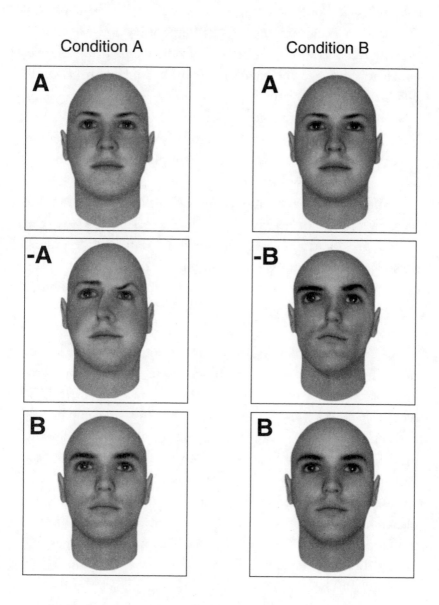

Note: For this illustration, I used computerized faces, not those of the MIT students. And of course, the letters did not appear on the original sheets.

nary students. But do you remember how the existence of a colonial-style house needing a new roof might push you to choose a perfect colonial over a contemporary house—simply because the decoy colonial would give you something against which to compare the regular colonial? And in the *Economist*'s ad, didn't the print-only option for $125 push people to take the print-and-Internet option for $125? Similarly, would the existence of a less perfect person (−A or −B) push people to choose the perfect one (A or B), simply because the decoy option served as a point of comparison?

It did. Whenever I handed out a sheet that had a regular picture, its inferior version, and another regular picture, the participants said they would prefer to date the "regular" person—the one who was similar, but clearly superior, to the distorted version—over the other, undistorted person on the sheet. This was not just a close call—it happened 75 percent of the time.

To explain the decoy effect further, let me tell you something about bread-making machines. When Williams-Sonoma first introduced a home "bread bakery" machine (for $275), most consumers were not interested. What was a home bread-making machine, anyway? Was it good or bad? Did one really need home-baked bread? Why not just buy a fancy coffee-maker sitting nearby instead? Flustered by poor sales, the manufacturer of the bread machine brought in a marketing research firm, which suggested a fix: introduce an additional model of the bread maker, one that was not only larger but priced about 50 percent higher than the initial machine.

Now sales began to rise (along with many loaves of bread), though it was not the large bread maker that was being sold. Why? Simply because consumers now had two models of bread makers to choose from. Since one was clearly larger and much more expensive than the other, people didn't have to make their decision in a vacuum. They could say: "Well, I don't know much about bread makers, but I do know that if I were to buy one, I'd rather have the smaller one for less money." And that's when bread makers began to fly off the shelves.[5]

OK for bread makers. But let's take a look at the decoy effect in a completely different situation. What if you are single, and hope to appeal to as many attractive potential dating partners as possible at an upcoming singles event? My advice would be to bring a friend who has your basic physical characteristics (similar coloring, body type, facial features), but is slightly less attractive (−you).

Why? Because the folks you want to attract will have a hard time evaluating you with no comparables around. However, if you are compared with a "−you," the decoy friend will do a lot to make you look better, not just in comparison with the decoy but also in general, and in comparison with all the other people around. It may sound irrational (and I can't guarantee this), but the chances are good that you will get some extra attention. Of course, don't just stop at looks. If great conversation will win the day, be sure to pick a friend for the singles event who can't match your smooth delivery and rapier wit. By comparison, you'll sound great.

Now that you know this secret, be careful: when a similar but better-looking friend of the same sex asks you to accompany him or her for a night out, you might wonder whether you have been invited along for your company or merely as a decoy.

RELATIVITY HELPS US make decisions in life. But it can also make us downright miserable. Why? Because jealousy and envy spring from comparing our lot in life with that of others.

It was for good reason, after all, that the Ten Commandments admonished, "Neither shall you desire your neighbor's
house nor field, or male or female slave, or donkey or anything that belongs to your neighbor." This might just be the
toughest commandment to follow, considering that by our
very nature we are wired to compare.

Modern life makes this weakness even more pronounced.
A few years ago, for instance, I met with one of the top executives of one of the big investment companies. Over the course
of our conversation he mentioned that one of his employees
had recently come to him to complain about his salary.

"How long have you been with the firm?" the executive
asked the young man.

"Three years. I came straight from college," was the
answer.

"And when you joined us, how much did you expect to be
making in three years?"

"I was hoping to be making about a hundred thousand."

The executive eyed him curiously.

"And now you are making almost three hundred thousand, so how can you possibly complain?" he asked.

"Well," the young man stammered, "it's just that a couple
of the guys at the desks next to me, they're not any better
than I am, and they are making three hundred ten."

The executive shook his head.

An ironic aspect of this story is that in 1993, federal securities regulators forced companies, for the first time, to reveal
details about the pay and perks of their top executives. The
idea was that once pay was in the open, boards would be reluctant to give executives outrageous salaries and benefits.
This, it was hoped, would stop the rise in executive compensation, which neither regulation, legislation, nor shareholder

pressure had been able to stop. And indeed, it needed to stop: in 1976 the average CEO was paid 36 times as much as the average worker. By 1993, the average CEO was paid 131 times as much.

But guess what happened. Once salaries became public information, the media regularly ran special stories ranking CEOs by pay. Rather than suppressing the executive perks, the publicity had CEOs in America comparing their pay with that of everyone else. In response, executives' salaries skyrocketed. The trend was further "helped" by compensation consulting firms (scathingly dubbed "Ratchet, Ratchet, and Bingo" by the investor Warren Buffett) that advised their CEO clients to demand outrageous raises. The result? Now the average CEO makes about 369 times as much as the average worker—about three times the salary before executive compensation went public.

Keeping that in mind, I had a few questions for the executive I met with.

"What would happen," I ventured, "if the information in your salary database became known throughout the company?"

The executive looked at me with alarm. "We could get over a lot of things here—insider trading, financial scandals, and the like—but if everyone knew everyone else's salary, it would be a true catastrophe. All but the highest-paid individual would feel underpaid—and I wouldn't be surprised if they went out and looked for another job."

Isn't this odd? It has been shown repeatedly that the link between amount of salary and happiness is not as strong as one would expect it to be (in fact, it is rather weak). Studies even find that countries with the "happiest" people are not among those with the highest personal income. Yet we keep pushing toward a higher salary. Much of that can be blamed on sheer envy. As H. L. Mencken, the twentieth-century journalist, satirist, social critic, cynic, and freethinker noted, a man's satisfaction with his salary depends on (are you ready for this?) whether he makes more than his wife's sister's husband. Why the wife's sister's husband? Because (and I have a feeling that Mencken's wife kept him fully informed of her sister's husband's salary) this is a comparison that is salient and readily available.*

All this extravagance in CEOs' pay has had a damaging effect on society. Instead of causing shame, every new outrage in compensation encourages other CEOs to demand even more. "In the Web World," according to a headline in the *New York Times*, the "Rich Now Envy the Superrich."

In another news story, a physician explained that he had graduated from Harvard with the dream of someday receiving a Nobel Prize for cancer research. This was his goal. This was his dream. But a few years later, he realized that several of his colleagues were making more as medical investment advisers at Wall Street firms than he was making in medicine. He had previously been happy with his income, but hearing of his friends' yachts and vacation homes, he suddenly felt very poor. So he took another route with his career—the route of Wall Street.[6] By the time he arrived at his twentieth class reunion, he was making 10 times what most of his peers were making in medicine. You can almost see him, standing in the middle of the room at the reunion, drink in hand—a large circle of influence with smaller circles gathering around him. He had not won the Nobel Prize, but

*Now that you know this fact, and assuming that you are not married, take this into account when you search for a soul mate. Look for someone whose sibling is married to a productivity-challenged individual.

he had relinquished his dreams for a Wall Street salary, for a chance to stop feeling "poor." Is it any wonder that family practice physicians, who make an average of $160,000 a year, are in short supply?*

CAN WE DO anything about this problem of relativity?

The good news is that we can sometimes control the "circles" around us, moving toward smaller circles that boost our relative happiness. If we are at our class reunion, and there's a "big circle" in the middle of the room with a drink in his hand, boasting of his big salary, we can consciously take several steps away and talk with someone else. If we are thinking of buying a new house, we can be selective about the open houses we go to, skipping the houses that are above our means. If we are thinking about buying a new car, we can focus on the models that we can afford, and so on.

We can also change our focus from narrow to broad. Let me explain with an example from a study conducted by two brilliant researchers, Amos Tversky and Daniel Kahneman. Suppose you have two errands to run today. The first is to buy a new pen, and the second is to buy a suit for work. At an office supply store, you find a nice pen for $25. You are set to buy it, when you remember that the same pen is on sale for $18 at another store 15 minutes away. What would you do? Do you decide to take the 15-minute trip to save the $7? Most people faced with this dilemma say that they would take the trip to save the $7.

Now you are on your second task: you're shopping for

---

*Of course, physicians have other problems as well, including insurance forms, bureaucracy, and threats of lawsuits for malpractice.

your suit. You find a luxurious gray pinstripe suit for $455 and decide to buy it, but then another customer whispers in your ear that the exact same suit is on sale for only $448 at another store, just 15 minutes away. Do you make this second 15-minute trip? In this case, most people say that they would not.

But what is going on here? Is 15 minutes of your time worth $7, or isn't it? In reality, of course, $7 is $7—no matter how you count it. The only question you should ask yourself in these cases is whether the trip across town, and the 15 extra minutes it would take, is worth the extra $7 you would save. Whether the amount from which this $7 will be saved is $10 or $10,000 should be irrelevant.

This is the problem of relativity—we look at our decisions in a relative way and compare them locally to the available alternative. We compare the relative advantage of the cheap pen with the expensive one, and this contrast makes it obvious to us that we should spend the extra time to save the $7. At the same time, the relative advantage of the cheaper suit is very small, so we spend the extra $7.

This is also why it is so easy for a person to add $200 to a $5,000 catering bill for a soup entrée, when the same person will clip coupons to save 25 cents on a one-dollar can of condensed soup. Similarly, we find it easy to spend $3,000 to upgrade to leather seats when we buy a new $25,000 car, but difficult to spend the same amount on a new leather sofa (even though we know we will spend more time at home on the sofa than in the car). Yet if we just thought about this in a broader perspective, we could better assess what we could do with the $3,000 that we are considering spending on upgrading the car seats. Would we perhaps be better off spending it on books, clothes, or a vacation? Thinking broadly like this is not easy, because making relative judgments is the natural way we think. Can you get a handle on it? I know someone who can.

He is James Hong, cofounder of the Hotornot.com rating and dating site. (James, his business partner Jim Young, Leonard Lee, George Loewenstein, and I recently worked on a research project examining how one's own "attractiveness" affects one's view of the "attractiveness" of others.)

For sure, James has made a lot of money, and he sees even more money all around him. One of his good friends, in fact, is a founder of PayPal and is worth tens of millions. But Hong knows how to make the circles of comparison in his life smaller, not larger. In his case, he started by selling his Porsche Boxster and buying a Toyota Prius in its place.[7]

"I don't want to live the life of a Boxster," he told the *New York Times*, "because when you get a Boxster you wish you had a 911, and you know what people who have 911s wish they had? They wish they had a Ferrari."

That's a lesson we can all learn: the more we have, the more we want. And the only cure is to break the cycle of relativity.

CHAPTER 2

The Fallacy of Supply
and Demand

Why the Price of Pearls—and Everything Else—
Is Up in the Air

At the onset of World War II, an Italian diamond dealer, James Assael, fled Europe for Cuba. There, he found a new livelihood: the American army needed waterproof watches, and Assael, through his contacts in Switzerland, was able to fill the demand.

When the war ended, Assael's deal with the U.S. government dried up, and he was left with thousands of Swiss watches. The Japanese needed watches, of course. But they didn't have any money. They did have pearls, though—many thousands of them. Before long, Assael had taught his son how to barter Swiss watches for Japanese pearls. The business blossomed, and shortly thereafter, the son, Salvador Assael, became known as the "pearl king."

The pearl king had moored his yacht at Saint-Tropez one day in 1973, when a dashing young Frenchman, Jean-Claude

Brouillet, came aboard from an adjacent yacht. Brouillet had just sold his air-freight business and with the proceeds had purchased an atoll in French Polynesia—a blue-lagooned paradise for himself and his young Tahitian wife. Brouillet explained that its turquoise waters abounded with black-lipped oysters, *Pinctada margaritifera*. And from the black lips of those oysters came something of note: black pearls.

At the time there was no market for Tahitian black pearls, and little demand. But Brouillet persuaded Assael to go into business with him. Together they would harvest black pearls and sell them to the world. At first, Assael's marketing efforts failed. The pearls were gunmetal gray, about the size of musket balls, and he returned to Polynesia without having made a single sale. Assael could have dropped the black pearls altogether or sold them at a low price to a discount store. He could have tried to push them to consumers by bundling them together with a few white pearls. But instead Assael waited a year, until the operation had produced some better specimens, and then brought them to an old friend, Harry Winston, the legendary gemstone dealer. Winston agreed to put them in the window of his store on Fifth Avenue, with an outrageously high price tag attached. Assael, meanwhile, commissioned a full-page advertisement that ran in the glossiest of magazines. There, a string of Tahitian black pearls glowed, set among a spray of diamonds, rubies, and emeralds.

The pearls, which had shortly before been the private business of a cluster of black-lipped oysters, hanging on a rope in the Polynesian sea, were soon parading through Manhattan on the arched necks of the city's most prosperous divas. Assael had taken something of dubious worth and made it fabulously fine. Or, as Mark Twain once noted about Tom

Sawyer, "Tom had discovered a great law of human action, namely, that in order to make a man covet a thing, it is only necessary to make the thing difficult to attain."

How did the pearl king do it? How did he persuade the cream of society to become passionate about Tahitian black pearls—and pay him royally for them? In order to answer this question, I need to explain something about baby geese.

A few decades ago, the naturalist Konrad Lorenz discovered that goslings, upon breaking out of their eggs, become attached to the first moving object they encounter (which is generally their mother). Lorenz knew this because in one experiment *he* became the first thing they saw, and they followed him loyally from then on through adolescence. With that, Lorenz demonstrated not only that goslings make initial decisions based on what's available in their environment, but that they stick with a decision once it has been made. Lorenz called this natural phenomenon *imprinting*.

Is the human brain, then, wired like that of a gosling? Do our first impressions and decisions become imprinted? And if so, how does this imprinting play out in our lives? When we encounter a new product, for instance, do we accept the first price that comes before our eyes? And more importantly, does that price (which in academic lingo we call an *anchor*) have a long-term effect on our willingness to pay for the product from then on?

It seems that what's good for the goose is good for humans as well. And this includes anchoring. From the beginning, for instance, Assael "anchored" his pearls to the finest gems in the world—and the prices followed forever after. Similarly, once we buy a new product at a particular price, we become anchored to that price. But how exactly does this work? Why do we accept anchors?

Consider this: if I asked you for the last two digits of your social security number (mine are 79), then asked you whether you would pay this number in dollars (for me this would be $79) for a particular bottle of Côtes du Rhône 1998, would the mere suggestion of that number influence how much you would be willing to spend on wine? Sounds preposterous, doesn't it? Well, wait until you see what happened to a group of MBA students at MIT a few years ago.

"NOW HERE WE have a nice Côtes du Rhône Jaboulet Parallel," said Drazen Prelec, a professor at MIT's Sloan School of Management, as he lifted a bottle admiringly. "It's a 1998."

At the time, sitting before him were the 55 students from his marketing research class. On this day, Drazen, George Loewenstein (a professor at Carnegie Mellon University), and I would have an unusual request for this group of future marketing pros. We would ask them to jot down the last two digits of their social security numbers and tell us whether they would pay this amount for a number of products, including the bottle of wine. Then, we would ask them to actually bid on these items in an auction.

What were we trying to prove? The existence of what we called *arbitrary coherence*. The basic idea of arbitrary coherence is this: although initial prices (such as the price of Assael's pearls) are "arbitrary," once those prices are established in our minds they will shape not only present prices but also future prices (this makes them "coherent"). So, would thinking about one's social security number be enough to create an anchor? And would that initial anchor have a long-term influence? That's what we wanted to see.

"For those of you who don't know much about wines," Drazen continued, "this bottle received eighty-six points from *Wine Spectator*. It has the flavor of red berry, mocha, and black chocolate; it's a medium-bodied, medium-intensity, nicely balanced red, and it makes for delightful drinking."

Drazen held up another bottle. This was a Hermitage Jaboulet La Chapelle, 1996, with a 92-point rating from the *Wine Advocate* magazine. "The finest La Chapelle since 1990," Drazen intoned, while the students looked up curiously. "Only 8,100 cases made . . ."

In turn, Drazen held up four other items: a cordless trackball (TrackMan Marble FX by Logitech); a cordless keyboard and mouse (iTouch by Logitech); a design book (*The Perfect Package: How to Add Value through Graphic Design*); and a one-pound box of Belgian chocolates by Neuhaus.

Drazen passed out forms that listed all the items. "Now I want you to write the last two digits of your social security number at the top of the page," he instructed. "And then write them again next to each of the items in the form of a price. In other words, if the last two digits are twenty-three, write twenty-three dollars."

"Now when you're finished with that," he added, "I want you to indicate on your sheets—with a simple yes or no— whether you would pay that amount for each of the products."

When the students had finished answering yes or no to each item, Drazen asked them to write down the maximum amount they were willing to pay for each of the products (their bids). Once they had written down their bids, the students passed the sheets up to me and I entered their responses into my laptop and announced the winners. One by one the student who had made the highest bid for each of the products would step up to the front of the class, pay for the product,* and take it with them.

The students enjoyed this class exercise, but when I asked them if they felt that writing down the last two digits of their social security numbers had influenced their final bids, they quickly dismissed my suggestion. No way!

When I got back to my office, I analyzed the data. Did the digits from the social security numbers serve as anchors? Remarkably, they did: the students with the highest-ending social security digits (from 80 to 99) bid highest, while those with the lowest-ending numbers (1 to 20) bid lowest. The top 20 percent, for instance, bid an average of $56 for the cordless keyboard; the bottom 20 percent bid an average of $16. In the end, we could see that students with social security numbers ending in the upper 20 percent placed bids that were 216 to 346 percent higher than those of the students with social security numbers ending in the lowest 20 percent (see table on the facing page).

Now if the last two digits of your social security number are a high number I know what you must be thinking: "I've been paying too much for everything my entire life!" This is not the case, however. Social security numbers were the anchor in this experiment only because we requested them. We could have just as well asked for the current temperature or the manufacturer's suggested retail price (MSRP). Any question, in fact, would have created the anchor. Does that seem rational? Of course not. But that's the way we are—goslings, after all.[†]

---

*The price the highest bidder paid for an item was based not on his own bid, but on that of the second highest bidder. This is called a second price auction. William Vickrey received the Nobel prize in economics for demonstrating that this type of auction creates the conditions where it is in people's best interest to bid the maximum amount they are willing to pay for each item (this is also the general logic behind the auction system on eBay).

†When I've tried this kind of experiment on executives and managers (at the MIT Execu-

Average prices paid for the various products for each of the five groups of final digits in social security numbers, and the correlations between these digits and the bids submitted in the auction.

Products	Range of last two digits of SS number					
	00–19	20–39	40–59	60–79	80–99	Correlations*
Cordless trackball	$8.64	$11.82	$13.45	$21.18	$26.18	0.42
Cordless keyboard	$16.09	$26.82	$29.27	$34.55	$55.64	0.52
Design book	$12.82	$16.18	$15.82	$19.27	$30.00	0.32
Neuhaus chocolates	$9.55	$10.64	$12.45	$13.27	$20.64	0.42
1998 Côtes du Rhône	$8.64	$14.45	$12.55	$15.45	$27.91	0.33
1996 Hermitage	$11.73	$22.45	$18.09	$24.55	$37.55	0.33

*Correlation is a statistical measure of how much the movement of two variables is related. The range of possible correlations is between –1 and +1, where a correlation of 0 means that the change in value of one variable has no bearing on the change in value of the other variable.

The data had one more interesting aspect. Although the willingness to pay for these items was arbitrary, there was also a logical, coherent aspect to it. When we looked at the bids for the two pairs of related items (the two wines and the two computer components), their relative prices seemed incredibly logical. Everyone was willing to pay more for the keyboard than for the trackball—and also pay more for the 1996 Hermitage than for the 1998 Côtes du Rhône. The significance of this is that once the participants were willing to pay a certain price for one product, their willingness to pay for other items in the same product category was judged relative to that first price (the anchor).

---

tive Education Program), I've had similar success making their social security numbers influence the prices they were willing to pay for chocolates, books, and other products.

This, then, is what we call arbitrary coherence. Initial prices are largely "arbitrary" and can be influenced by responses to random questions; but once those prices are established in our minds, they shape not only what we are willing to pay for an item, but also how much we are willing to pay for related products (this makes them coherent).

Now I need to add one important clarification to the story I've just told. In life we are bombarded by prices. We see the manufacturer's suggested retail price (MSRP) for cars, lawn mowers, and coffeemakers. We get the real estate agent's spiel on local housing prices. But price tags by themselves are not necessarily anchors. They become anchors when we contemplate buying a product or service at that particular price. That's when the imprint is set. From then on, we are willing to accept a range of prices—but as with the pull of a bungee cord, we always refer back to the original anchor. Thus the first anchor influences not only the immediate buying decision but many others that follow.

We might see a 57-inch LCD high-definition television on sale for $3,000, for instance. The price tag is not the anchor. But if we decide to buy it (or seriously contemplate buying it) at that price, then the decision becomes our anchor henceforth in terms of LCD television sets. That's our peg in the ground, and from then on—whether we shop for another set or merely have a conversation at a backyard cookout—all other high-definition televisions are judged relative to that price.

Anchoring influences all kinds of purchases. Uri Simonsohn (a professor at the University of Pennsylvania) and George Loewenstein, for example, found that people who move to a new city generally remain anchored to the prices they paid for housing in their former city. In their study they found that people who move from inexpensive markets (say, Lubbock,

Texas) to moderately priced cities (say, Pittsburgh) don't increase their spending to fit the new market.* Rather, these people spend an amount similar to what they were used to in the previous market, even if this means having to squeeze themselves and their families into smaller or less comfortable homes. Likewise, transplants from more expensive cities sink the same dollars into their new housing situation as they did in the past. People who move from Los Angeles to Pittsburgh, in other words, don't generally downsize their spending much once they hit Pennsylvania: they spend an amount similar to what they used to spend in Los Angeles.

It seems that we get used to the particularities of our housing markets and don't readily change. The only way out of this box, in fact, is to rent a home in the new location for a year or so. That way, we adjust to the new environment—and, after a while, we are able to make a purchase that aligns with the local market.

So we anchor ourselves to initial prices. But do we hop from one anchor price to another (flip-flopping, if you will), continually changing our willingness to pay? Or does the first anchor we encounter become our anchor for a long time and for many decisions? To answer this question, we decided to conduct another experiment—one in which we attempted to lure our participants from old anchors to new ones.

For this experiment we enlisted some undergraduate students, some graduate students, and some investment bankers who had come to the campus to recruit new employees for their firms. Once the experiment started we presented our

*The result was not due to wealth, taxes, or other financial reasons.

31

participants with three different sounds, and following each, asked them if they would be willing to get paid a particular amount of money (which served as the price anchor) for hearing those sounds again. One sound was a 30-second high-pitched 3,000-hertz sound, somewhat like someone screaming in a high-pitched voice. Another was a 30-second full-spectrum noise (also called white noise), which is similar to the noise a television set makes when there is no reception. The third was a 30-second oscillation between high-pitched and low-pitched sounds. (I am not sure if the bankers understood exactly what they were about to experience, but maybe even our annoying sounds were less annoying than talking about investment banking.)

We used sounds because there is no existing market for annoying sounds (so the participants couldn't use a market price as a way to think about the value of these sounds). We also used annoying sounds, specifically, because no one likes such sounds (if we had used classical music, some would have liked it better than others). As for the sounds themselves, I selected them after creating hundreds of sounds, choosing these three because they were, in my opinion, equally annoying.

We placed our participants in front of computer screens at the lab, and had them clamp headphones over their ears.

As the room quieted down, the first group saw this message appear in front of them: "In a few moments we are going to play a new unpleasant tone over your headset. We are interested in how annoying you find it. Immediately after you hear the tone, we will ask you whether, hypothetically, you would be willing to repeat the same experience in exchange for a payment of 10 cents." The second group got the same message, only with an offer of 90 cents rather than 10 cents.

Would the anchor prices make a difference? To find out,

we turned on the sound—in this case the irritating 30-second, 3,000-hertz squeal. Some of our participants grimaced. Others rolled their eyes.

When the screeching ended, each participant was presented with the anchoring question, phrased as a hypothetical choice: Would the participant be willing, hypothetically, to repeat the experience for a cash payment (which was 10 cents for the first group and 90 cents for the second group)? After answering this anchoring question, the participants were asked to indicate on the computer screen the lowest price they would demand to listen to the sound again. This decision was real, by the way, as it would determine whether they would hear the sound again—and get paid for doing so.*

Soon after the participants entered their prices, they learned the outcome. Participants whose price was sufficiently low "won" the sound, had the (unpleasant) opportunity to hear it again, and got paid for doing so. The participants whose price was too high did not listen to the sound and were not paid for this part of the experiment.

What was the point of all this? We wanted to find out whether the first prices that we suggested (10 cents and 90 cents) had served as an anchor. And indeed they had. Those who first faced the hypothetical decision about whether to listen to the sound for 10 cents needed much less money to be willing to listen to this sound again (33 cents on average) relative to those who first faced the hypothetical decision about whether to listen to the sound for 90 cents—this second group demanded more than twice the compensation (73

---

*To ensure that the bids we got were indeed the lowest prices for which the participants would listen to the annoying sounds, we used the "Becker-DeGroot-Marschak procedure." This is an auction-like procedure, in which each of the participants bids against a price randomly drawn by a computer.

cents on average) for the same annoying experience. Do you see the difference that the suggested price had?

BUT THIS WAS only the start of our exploration. We also wanted to know how influential the anchor would be in future decisions. Suppose we gave the participants an opportunity to drop this anchor and run for another? Would they do it? To put it in terms of goslings, would they swim across the pond after their original imprint and then, midway, swing their allegiance to a new mother goose? In terms of goslings, I think you know that they would stick with the original mom. But what about humans? The next two phases of the experiment would enable us to answer these questions.

In the second phase of the experiment, we took participants from the previous 10-cents and 90-cents groups and treated them to 30 seconds of a white, wooshing noise. "Hypothetically, would you listen to this sound again for 50 cents?" we asked them at the end. The respondents pressed a button on their computers to indicate yes or no.

"OK, *how much* would you need to be paid for this?" we asked. Our participants typed in their lowest price; the computer did its thing; and, depending on their bids, some participants listened to the sound again and got paid and some did not. When we compared the prices, the 10-cents group offered much lower bids than the 90-cents group. This means that although both groups had been equally exposed to the suggested 50 cents, as their focal anchoring response (to "Hypothetically, would you listen to this sound again for 50 cents?"), the first anchor in this annoying sound category (which was 10 cents for some and 90 cents for others) predominated.

Why? Perhaps the participants in the 10-cents group said

something like the following to themselves: "Well, I listened previously to that annoying sound for a low amount. This sound is not much different. So if I said a low amount for the previous one, I guess I could bear this sound for about the same price." Those who were in the 90-cents group used the same type of logic, but because their starting point was different, so was their ending point. These individuals told themselves, "Well, I listened previously to that annoying sound for a high amount. This sound is not much different. So since I said a high amount for the previous one, I guess I could bear this sound for about the same price." Indeed, the effect of the first anchor held—indicating that anchors have an enduring effect for present prices as well as for future prices.

There was one more step to this experiment. This time we had our participants listen to the oscillating sound that rose and fell in pitch for 30 seconds. We asked our 10-cents group, "Hypothetically, would you listen to this sound again for 90 cents?" Then we asked our 90-cents group, "Would you listen to this sound again for 10 cents?" Having flipped our anchors, we would now see which one, the local anchor or the first anchor, exerted the greatest influence.

Once again, the participants typed in yes or no. Then we asked them for real bids: "How much would it take for you to listen to this again?" At this point, they had a history with three anchors: the first one they encountered in the experiment (either 10 cents or 90 cents), the second one (50 cents), and the most recent one (either 90 cents or 10 cents). Which one of these would have the largest influence on the price they demanded to listen to the sound?

Again, it was as if our participants' minds told them, "If I listened to the first sound for x cents, and listened to the second sound for *x* cents as well, then I can surely do this one for *x* cents, too!" And that's what they did. Those who had first encountered the 10-cent anchor accepted low prices, even after 90 cents was suggested as the anchor. On the other hand, those who had first encountered the 90-cent anchor kept on demanding much higher prices, regardless of the anchors that followed.

What did we show? That our first decisions resonate over a long sequence of decisions. First impressions are important, whether they involve remembering that our first DVD player cost much more than such players cost today (and realizing that, in comparison, the current prices are a steal) or remembering that gas was once a dollar a gallon, which makes every trip to the gas station a painful experience. In all these cases the random, and not so random, anchors that we encountered along the way and were swayed by remain with us long after the initial decision itself.

NOW THAT WE know we behave like goslings, it is important to understand the process by which our first decisions translate into long-term habits. To illustrate this process, consider this example. You're walking past a restaurant, and you see two people standing in line, waiting to get in. "This must be a good restaurant," you think to yourself. "People are standing in line." So you stand behind these people. Another person walks by. He sees three people standing in line and thinks, "This must be a fantastic restaurant," and joins the line. Others join. We call this type of behavior herding. It happens when we assume that something is good (or bad) on the basis of other people's previous behavior, and our own actions follow suit.

But there's also another kind of herding, one that we call self-herding. This happens when we believe something is good (or bad) on the basis of our own previous behavior. Essentially, once we become the first person in line at the restaurant, we begin to line up behind ourself in subsequent experiences. Does that make sense? Let me explain.

Recall your first introduction to Starbucks, perhaps several years ago. (I assume that nearly everyone has had this experience, since Starbucks sits on every corner in America.) You are sleepy and in desperate need of a liquid energy boost as you embark on an errand one afternoon. You glance through the windows at Starbucks and walk in. The prices of the coffee are a shock—you've been blissfully drinking the brew at Dunkin' Donuts for years. But since you have walked in and are now curious about what coffee at this price might taste like, you surprise yourself: you buy a small coffee, enjoy its taste and its effect on you, and walk out.

The following week you walk by Starbucks again. Should you go in? The ideal decision-making process should take into account the quality of the coffee (Starbucks versus Dunkin' Donuts); the prices at the two places; and, of course, the cost (or value) of walking a few more blocks to get to Dunkin' Donuts. This is a complex computation—so instead, you resort to the simple approach: "I went to Starbucks before, and I enjoyed myself and the coffee, so this must be a good decision for me." So you walk in and get another small cup of coffee.

In doing so, you just became the second person in line, standing behind yourself. A few days later, you again walk by Starbucks and this time, you vividly remember your past decisions and act on them again—voilà! You become the third person in line, standing behind yourself. As the weeks pass, you enter again and again and every time, you feel more strongly that you are acting on the basis of your preferences. Buying coffee at Starbucks has become a habit with you.

BUT THE STORY doesn't end there. Now that you have gotten used to paying more for coffee, and have bumped yourself up onto a new curve of consumption, other changes also become simpler. Perhaps you will now move up from the small cup for $2.20 to the medium size for $3.50 or to the Venti for $4.15. Even though you don't know how you got into this price bracket in the first place, moving to a larger coffee at a relatively greater price seems pretty logical. So is a lateral move to other offerings at Starbucks: Caffè Americano, Caffè Misto, Macchiato, and Frappuccino, for instance.

If you stopped to think about this, it would not be clear whether you should be spending all this money on coffee at Starbucks instead of getting cheaper coffee at Dunkin' Donuts or even free coffee at the office. But you don't think about these trade-offs anymore. You've already made this decision many times in the past, so you now assume that this is the way you want to spend your money. You've herded yourself—lining up behind your initial experience at Starbucks—and now you're part of the crowd.

HOWEVER, THERE IS something odd in this story. If anchoring is based on our initial decisions, how did Starbucks manage to become an initial decision in the first place? In other words, if we were previously anchored to the prices at Dunkin' Donuts, how did we move our anchor to Starbucks? This is where it gets really interesting.

When Howard Shultz created Starbucks, he was as intuitive a businessman as Salvador Assael. He worked diligently to separate Starbucks from other coffee shops, not through price but through ambience. Accordingly, he designed Starbucks from the very beginning to feel like a continental coffeehouse.

The early shops were fragrant with the smell of roasted beans (and better-quality roasted beans than those at Dunkin' Donuts). They sold fancy French coffee presses. The showcases presented alluring snacks—almond croissants, biscotti, raspberry custard pastries, and others. Whereas Dunkin' Donuts had small, medium, and large coffees, Starbucks offered Short, Tall, Grande, and Venti, as well as drinks with high-pedigree names like Caffè Americano, Caffè Misto, Macchiato, and Frappuccino. Starbucks did everything in its power, in other words, to make the experience feel different—so different that we would not use the prices at Dunkin' Donuts as an anchor, but instead would be open to the new anchor that Starbucks was preparing for us. And that, to a great extent, is how Starbucks succeeded.

GEORGE, DRAZEN, AND I were so excited with the experiments on coherent arbitrariness that we decided to push the idea one step farther. This time, we had a different twist to explore.

Do you remember the famous episode in *The Adventures of Tom Sawyer*, the one in which Tom turned the whitewashing of Aunt Polly's fence into an exercise in manipulating his friends? As I'm sure you recall, Tom applied the paint with gusto, pretending to enjoy the job. "Do you call this work?" Tom told his friends. "Does a boy get a chance to whitewash a fence every day?" Armed with this new "information," his friends discovered the joys of whitewashing a fence. Before long, Tom's friends were not only paying him for the privilege, but deriving real pleasure from the task—a win-win outcome if there ever was one.

From our perspective, Tom transformed a negative experience to a positive one—he transformed a situation in which compensation was required to one in which people (Tom's friends) would pay to get in on the fun. Could we do the same? We thought we'd give it a try.

One day, to the surprise of my students, I opened the day's lecture on managerial psychology with a poetry selection, a few lines of "Whoever you are holding me now in hand" from Walt Whitman's *Leaves of Grass*:

> *Whoever you are holding me now in hand,*
> *Without one thing all will be useless,*
> *I give you fair warning before you attempt me*
>   *further,*
> *I am not what you supposed, but far different.*
> *Who is he that would become my follower?*
> *Who would sign himself a candidate for my*
>   *affections?*
> *The way is suspicious, the result uncertain, perhaps*
>   *destructive,*
> *You would have to give up all else, I alone would*
>   *expect to be your sole and exclusive standard,*
> *Your novitiate would even then be long and*
>   *exhausting,*
> *The whole past theory of your life and all*
>   *conformity to the lives around you would have to*
>   *be abandon'd,*
> *Therefore release me now before troubling yourself*

any further, let go your hand from my shoulders,
Put me down and depart on your way.

After closing the book, I told the students that I would be conducting three readings from Walt Whitman's *Leaves of Grass* that Friday evening: one short, one medium, and one long. Owing to limited space, I told them, I had decided to hold an auction to determine who could attend. I passed out sheets of paper so that they could bid for a space; but before they did so, I had a question to ask them.

I asked half the students to write down whether, hypothetically, they would be willing to pay me $10 for a 10-minute poetry recitation. I asked the other half to write down whether, hypothetically, they would be willing to listen to me recite poetry for ten minutes if I paid them $10.

This, of course, served as the anchor. Now I asked the students to bid for a spot at my poetry reading. Do you think the initial anchor influenced the ensuing bids?

Before I tell you, consider two things. First, my skills at reading poetry are not of the first order. So asking someone to pay me for 10 minutes of it could be considered a stretch. Second, even though I asked half of the students if they would pay me for the privilege of attending the recitation, they didn't have to bid that way. They could have turned the tables completely and demanded that I pay them.

And now to the results (drumroll, please). Those who answered the hypothetical question about paying me were indeed willing to pay me for the privilege. They offered, on average, to pay me about a dollar for the short poetry reading, about two dollars for the medium poetry reading, and a bit more than three dollars for the long poetry reading. (Maybe I could make a living outside academe after all.)

But, what about those who were anchored to the thought of being paid (rather than paying me)? As you might expect, they demanded payment: on average, they wanted $1.30 to listen to the short poetry reading, $2.70 to listen to the medium poetry reading, and $4.80 to endure the long poetry reading.

Much like Tom Sawyer, then, I was able to take an ambiguous experience (and if you could hear me recite poetry, you would understand just how ambiguous this experience is) and arbitrarily make it into a pleasurable or painful experience. Neither group of students knew whether my poetry reading was of the quality that is worth paying for or of the quality that is worth listening to only if one is being financially compensated for the experience (they did not know if it is pleasurable or painful). But once the first impression had been formed (that they would pay me or that I would pay them), the die was cast and the anchor set. Moreover, once the first decision had been made, other decisions followed in what seemed to be a logical and coherent manner. The students did not know whether listening to me recite poetry was a good or bad experience, but whatever their first decision was, they used it as input for their subsequent decisions and provided a coherent pattern of responses across the three poetry readings.

Of course, Mark Twain came to the same conclusions: "If Tom had been a great and wise philosopher, like the writer of this book, he would now have comprehended that work consists of whatever a body is obliged to do, and that play consists of whatever a body is not obliged to do." Mark Twain further observed: "There are wealthy gentlemen in England who drive four-horse passenger-coaches twenty or thirty miles on a daily line in the summer because the privilege costs them considerable money; but if they were offered wages for the service, that would turn it into work, and then they would resign."*

WHERE DO THESE thoughts lead us? For one, they illustrate the many choices we make, from the trivial to the profound, in which anchoring plays a role. We decide whether or not to purchase Big Macs, smoke, run red lights, take vacations in Patagonia, listen to Tchaikovsky, slave away at doctoral dissertations, marry, have children, live in the suburbs, vote Republican, and so on. According to economic theory, we base these decisions on our fundamental values—our likes and dislikes.

But what are the main lessons from these experiments about our lives in general? Could it be that the lives we have so carefully crafted are largely just a product of arbitrary coherence? Could it be that we made arbitrary decisions at some point in the past (like the goslings that adopted Lorenz as their parent) and have built our lives on them ever since, assuming that the original decisions were wise? Is that how we chose our careers, our spouses, the clothes we wear, and the way we style our hair? Were they smart decisions in the first place? Or were they partially random first imprints that have run wild?

Descartes said, *Cogito ergo sum*—"I think, therefore I am." But suppose we are nothing more than the sum of our first, naive, random behaviors. What then?

These questions may be tough nuts to crack, but in terms of our personal lives, we can actively improve on our irrational

---

*We will return to this astute observation in the chapter on social and market norms (Chapter 4).

behaviors. We can start by becoming aware of our vulnerabilities. Suppose you're planning to buy a cutting-edge cell phone (the one with the three-megapixel, 8× zoom digital camera), or even a daily $4 cup of gourmet coffee. You might begin by questioning that habit. How did it begin? Second, ask yourself what amount of pleasure you will be getting out of it. Is the pleasure as much as you thought you would get? Could you cut back a little and better spend the remaining money on something else? With everything you do, in fact, you should train yourself to question your repeated behaviors. In the case of the cell phone, could you take a step back from the cutting edge, reduce your outlay, and use some of the money for something else? And as for the coffee—rather than asking which blend of coffee you will have today, ask yourself whether you should even be having that habitual cup of expensive coffee at all.*

We should also pay particular attention to the first decision we make in what is going to be a long stream of decisions (about clothing, food, etc.). When we face such a decision, it might seem to us that this is just one decision, without large consequences; but in fact the power of the first decision can have such a long-lasting effect that it will percolate into our future decisions for years to come. Given this effect, the first decision is crucial, and we should give it an appropriate amount of attention.

Socrates said that the unexamined life is not worth living. Perhaps it's time to inventory the imprints and anchors in our own life. Even if they once were completely reasonable, are they still reasonable? Once the old choices are reconsidered,

---

*I am not claiming that spending money on a wonderful cup of coffee every day, or even a few times a day, is necessarily a bad decision—I am saying only that we should question our decisions.

we can open ourselves to new decisions—and the new opportunities of a new day. That seems to make sense.

ALL THIS TALK about anchors and goslings has larger implications than consumer preferences, however. Traditional economics assumes that prices of products in the market are determined by a balance between two forces: production at each price (supply) and the desires of those with purchasing power at each price (demand). The price at which these two forces meet determines the prices in the marketplace.

This is an elegant idea, but it depends centrally on the assumption that the two forces are independent and that together they produce the market price. The results of all the experiments presented in this chapter (and the basic idea of arbitrary coherence itself) challenge these assumptions. First, according to the standard economic framework, consumers' willingness to pay is one of the two inputs that determine market prices (this is the demand). But as our experiments demonstrate, what consumers are willing to pay can easily be manipulated, and this means that consumers don't in fact have a good handle on their own preferences and the prices they are willing to pay for different goods and experiences.

Second, whereas the standard economic framework assumes that the forces of supply and demand are independent, the type of anchoring manipulations we have shown here suggest that they are, in fact, dependent. In the real world, anchoring comes from manufacturer's suggested retail prices (MSRPs), advertised prices, promotions, product introductions, etc.—all of which are supply-side variables. It seems then that instead of consumers' willingness to pay influencing market prices, the causality is somewhat reversed and it is market prices themselves that influence consumers' willingness to pay. What this means is that demand is not, in fact, a completely separate force from supply.

AND THIS IS not the end of the story. In the framework of arbitrary coherence, the relationships we see in the marketplace between demand and supply (for example, buying more yogurt when it is discounted) are based not on preferences but on memory. Here is an illustration of this idea. Consider your current consumption of milk and wine. Now imagine that two new taxes will be introduced tomorrow. One will cut the price of wine by 50 percent, and the other will increase the price of milk by 100 percent. What do you think will happen? These price changes will surely affect consumption, and many people will walk around slightly happier and with less calcium. But now imagine this. What if the new taxes are accompanied by induced amnesia for the previous prices of wine and milk? What if the prices change in the same way, but you do not remember what you paid for these two products in the past?

I suspect that the price changes would make a huge impact on demand if people remembered the previous prices and noticed the price increases; but I also suspect that without a memory for past prices, these price changes would have a trivial effect, if any, on demand. If people had no memory of past prices, the consumption of milk and wine would remain essentially the same, as if the prices had not changed. In other words, the sensitivity we show to price changes might in fact be largely a result of our memory for the prices we have paid in the past and our desire for coherence with our past decisions—not at all a reflection of our true preferences or our level of demand.

The same basic principle would also apply if the government one day decided to impose a tax that doubled the price of gasoline. Under conventional economic theory, this should cut demand. But would it? Certainly, people would initially compare the new prices with their anchor, would be flabbergasted by the new prices, and so might pull back on their gasoline consumption and maybe even get a hybrid car. But over the long run, and once consumers readjusted to the new price and the new anchors (just as we adjust to the price of Nike sneakers, bottled water, and everything else), our gasoline consumption, at the new price, might in fact get close to the pretax level. Moreover, much as in the example of Starbucks, this process of readjustment could be accelerated if the price change were to also be accompanied by other changes, such as a new grade of gas, or a new type of fuel (such as corn-based ethanol fuel).

I am not suggesting that doubling the price of gasoline would have no effect on consumers' demand. But I do believe that in the long term, it would have a much smaller influence on demand than would be assumed from just observing the short-term market reactions to price increases.

ANOTHER IMPLICATION OF arbitrary coherence has to do with the claimed benefits of the free market and free trade. The basic idea of the free market is that if I have something that you value more than I do—let's say a sofa—trading this item will benefit both of us. This means that the mutual benefit of trading rests on the assumption that all the players in the market know the value of what they have and the value of the things they are considering getting from the trade.

But if our choices are often affected by random initial anchors, as we observed in our experiments, the choices and

trades we make are not necessarily going to be an accurate reflection of the real pleasure or utility we derive from those products. In other words, in many cases we make decisions in the
marketplace that may not reflect how much pleasure we can get
from different items. Now, if we can't accurately compute these
pleasure values, but frequently follow arbitrary anchors instead,
then it is not clear that the opportunity to trade is necessarily
going to make us better off. For example, because of some unfortunate initial anchors we might mistakenly trade something
that truly gives us a lot of pleasure (but regrettably had a low
initial anchor) for something that gives us less pleasure (but owing to some random circumstances had a high initial anchor). If
anchors and memories of these anchors—but not preferences—
determine our behavior, why would trading be hailed as the key
to maximizing personal happiness (utility)?

So, WHERE DOES this leave us? If we can't rely on the market
forces of supply and demand to set optimal market prices, and
we can't count on free-market mechanisms to help us maximize our utility, then we may need to look elsewhere. This is
especially the case with society's essentials, such as health care,
medicine, water, electricity, education, and other critical resources. If you accept the premise that market forces and free
markets will not always regulate the market for the best, then
you may find yourself among those who believe that the government (we hope a reasonable and thoughtful government)
must play a larger role in regulating some market activities,
even if this limits free enterprise. Yes, a free market based on
supply, demand, and no friction would be the ideal if we were
truly rational. Yet when we are not rational but irrational,
policies should take this important factor into account.

CHAPTER 3

The Cost of Zero Cost

Why We Often Pay Too Much When We Pay Nothing

Have you ever grabbed for a coupon offering a FREE! package of coffee beans—even though you don't drink coffee and don't even have a machine with which to brew it? What about all those FREE! extra helpings you piled on your plate at a buffet, even though your stomach had already started to ache from all the food you had consumed? And what about the worthless FREE! stuff you've accumulated— the promotional T-shirt from the radio station, the teddy bear that came with the box of Valentine chocolates, the magnetic calendar your insurance agent sends you each year?

It's no secret that getting something free feels very good. Zero is not just another price, it turns out. Zero is an emotional hot button—a source of irrational excitement. Would you buy something if it were discounted from 50 cents to 20 cents? Maybe. Would you buy it if it were discounted from 50 cents to two cents? Maybe. Would you grab it if it were discounted from 50 cents to zero? You bet!

What is it about zero cost that we find so irresistible? Why does FREE! make us so happy? After all, FREE! can lead us into trouble: things that we would never consider purchasing become incredibly appealing as soon as they are FREE! For instance, have you ever gathered up free pencils, key chains, and notepads at a conference, even though you'd have to carry them home and would only throw most of them away? Have you ever stood in line for a very long time (too long), just to get a free cone of Ben and Jerry's ice cream? Or have you bought two of a product that you wouldn't have chosen in the first place, just to get the third one for free?

ZERO HAS HAD a long history. The Babylonians invented the concept of zero; the ancient Greeks debated it in lofty terms (how could something be nothing?); the ancient Indian scholar Pingala paired zero with the numeral 1 to get double digits; and both the Mayans and the Romans made zero part of their numeral systems. But zero really found its place about AD 498, when the Indian astronomer Aryabhata sat up in bed one morning and exclaimed, *"Sthanam sthanam dasa gunam"*— which translates, roughly, as "Place to place in 10 times in value." With that, the idea of decimal-based place-value notation was born. Now zero was on a roll: It spread to the Arab world, where it flourished; crossed the Iberian Peninsula to Europe (thanks to the Spanish Moors); got some tweaking from the Italians; and eventually sailed the Atlantic to the New World, where zero ultimately found plenty of employment (together with the digit 1) in a place called Silicon Valley.

So much for a brief recounting of the history of zero. But the concept of zero applied to money is less clearly understood. In fact, I don't think it even has a history. Nonetheless, FREE!

has huge implications, extending not only to discount prices and promotions, but also to how FREE! can be used to help us make decisions that would benefit ourselves and society.

If FREE! were a virus or a subatomic particle, I might use an electron microscope to probe the object under the lens, stain it with different compounds to reveal its nature, or somehow slice it apart to reveal its inner composition. In behavioral economics we use a different instrument, however, one that allows us to slow down human behavior and examine it frame by frame, as it unfolds. As you have undoubtedly guessed by now, this procedure is called an experiment.

IN ONE EXPERIMENT, Kristina Shampanier (a PhD student at MIT), Nina Mazar (a professor at the University of Toronto), and I went into the chocolate business. Well, sort of. We set up a table at a large public building and offered two kinds of chocolates—Lindt truffles and Hershey's Kisses. There was a large sign above our table that read, "One chocolate per customer." Once the potential customers stepped closer, they could see the two types of chocolate and their prices.*

For those of you who are not chocolate connoisseurs, Lindt is produced by a Swiss firm that has been blending fine cocoas for 160 years. Lindt's chocolate truffles are particularly prized— exquisitely creamy and just about irresistible. They cost about 30 cents each when we buy them in bulk. Hershey's Kisses, on the other hand, are good little chocolates, but let's face it, they are rather ordinary: Hershey cranks out 80 million Kisses a day. In Hershey, Pennsylvania, even the streetlamps are made in the shape of the ubiquitous Hershey's Kiss.

---

*We posted the prices so that they were visible only when people got close to the table. We did this because we wanted to make sure that we did not attract different types of people in the different conditions—avoiding what is called self-selection.

So what happened when the "customers" flocked to our table? When we set the price of a Lindt truffle at 15 cents and a Kiss at one cent, we were not surprised to find that our customers acted with a good deal of rationality: they compared the price and quality of the Kiss with the price and quality of the truffle, and then made their choice. About 73 percent of them chose the truffle and 27 percent chose a Kiss.

Now we decided to see how FREE! might change the situation. So we offered the Lindt truffle for 14 cents and the Kisses free. Would there be a difference? Should there be? After all, we had merely lowered the price of both kinds of chocolate by one cent.

But what a difference FREE! made. The humble Hershey's Kiss became a big favorite. Some 69 percent of our customers (up from 27 percent before) chose the FREE! Kiss, giving up the opportunity to get the Lindt truffle for a very good price. Meanwhile, the Lindt truffle took a tumble; customers choosing it decreased from 73 to 31 percent.

What was going on here? First of all, let me say that there are many times when getting FREE! items can make perfect sense. If you find a bin of free athletic socks at a department store, for instance, there's no downside to grabbing all the socks you can. The critical issue arises when FREE! becomes a struggle between a free item and another item—a struggle in which the presence of FREE! leads us to make a bad decision. For instance, imagine going to a sports store to buy a pair of white socks, the kind with a nicely padded heel and a gold toe. Fifteen minutes later you're leaving the store, not with the socks you came in for, but with a cheaper pair that you don't like at all (without a padded heel and gold toe) but that came in a package with a FREE! second pair. This is a case in which you gave up a better deal and settled for something that was not what you wanted, just because you were lured by the FREE!

To replicate this experience in our chocolate experiment, we told our customers that they could choose only a single sweet—the Kiss or the truffle. It was an either-or decision, like choosing one kind of athletic sock over another. That's what made the customers' reaction to the FREE! Kiss so dramatic: Both chocolates were discounted by the same amount of money. The relative price difference between the two was unchanged—and so was the expected pleasure from both.

According to standard economic theory (simple cost-benefit analysis), then, the price reduction should not lead to any change in the behavior of our customers. Before, about 27 percent chose the Kiss and 73 percent chose the truffle. And since nothing had changed in relative terms, the response to the price reduction should have been exactly the same. A passing economist, twirling his cane and espousing conventional economic theory, in fact, would have said that since everything in the situation was the same, our customers should have chosen the truffles by the same margin of preference.*

And yet here we were, with people pressing up to the table to grab our Hershey's Kisses, not because they had made a reasoned cost-benefit analysis before elbowing their way in, but simply because the Kisses were FREE! How strange (but predictable) we humans are!

THIS CONCLUSION, INCIDENTALLY, remained the same in other experiments as well. In one case we priced the Hershey's Kiss at two cents, one cent, and zero cents, while pricing the truffle correspondingly at 27 cents, 26 cents, and 25 cents.

---

*For a more detailed account of how a rational consumer should make decisions in these cases, see the appendix to this chapter.

We did this to see if discounting the Kiss from two cents to one cent and the truffle from 27 cents to 26 cents would make a difference in the proportion of buyers for each. It didn't. But, once again, when we lowered the price of the Kiss to free, the reaction was dramatic. The shoppers overwhelmingly demanded the Kisses.

We decided that perhaps the experiment had been tainted, since shoppers may not feel like searching for change in a purse or backpack, or they may not have any money on them. Such an effect would artificially make the free offer seem more attractive. To address this possibility, we ran other experiments at one of MIT's cafeterias. In this setup, the chocolates were displayed next to the cashier as one of the cafeteria's regular promotions and the students who were interested in the chocolates simply added them to the lunch purchase, and paid for them while going through the cashier's line. What happened? The students still went overwhelmingly for the FREE! option.

WHAT IS IT about FREE! that's so enticing? Why do we have an irrational urge to jump for a FREE! item, even when it's not what we really want?

I believe the answer is this. Most transactions have an upside and a downside, but when something is FREE! we forget the downside. FREE! gives us such an emotional charge that we perceive what is being offered as immensely more valuable than it really is. Why? I think it's because humans are intrinsically afraid of loss. The real allure of FREE! is tied to this fear. There's no visible possibility of loss when we choose a FREE! item (it's free). But suppose we choose the item that's *not* free. Uh-oh, now there's a risk of having made a poor decision—the possibility of a loss. And so, given the choice, we go for what is free.

For this reason, in the land of pricing, zero is not just another price. Sure, 10 cents can make a huge difference in demand (suppose you were selling millions of barrels of oil), but nothing beats the emotional surge of FREE! This, the *zero price effect*, is in a category all its own.

To be sure, "buying something for nothing" is a bit of an oxymoron. But let me give you an example of how we often fall into the trap of buying something we may not want, simply because of that sticky substance, FREE!

In 2007, I saw a newspaper ad from a major electronics maker, offering me seven FREE! DVD titles if I purchased the maker's new high-definition DVD player. First of all, did I need a high-definition player at that time? Probably not. But even if I had, wouldn't it have been wiser to wait for prices to descend? They always do—and today's $600 high-definition DVD player will very quickly be tomorrow's $200 machine. Second, the DVD maker had a clear agenda behind its offer. This company's high-definition DVD system was in cutthroat competition with Blu-Ray, a system backed by many other manufacturers. At the time, Blu-Ray was ahead and has since gone on to dominate the market. So how much is FREE! when the machine being offered will find its way into obsolescence (like Betamax VCRs)? Those are two rational thoughts that might prevent us from falling under the spell of FREE! But, gee, those FREE! DVDs certainly look good!

GETTING SOMETHING FREE! is certainly a draw when we talk about prices. But what would happen if the offer was not a free price, but a free exchange? Are we as susceptible to free products as we are to getting products for free? A few years ago, with Halloween drawing near, I had an idea for an experiment to probe that question. This time I wouldn't even have to leave my home to get my answers.

Early in the evening, Joey, a nine-year-old kid dressed as Spider-Man and carrying a large yellow bag, climbed the stairs of our front porch. His mother accompanied him, to ensure that no one gave her kid an apple with a razor blade inside. (By the way, there never was a case of razor blades being distributed in apples on Halloween; it is just an urban myth.) She stayed on the sidewalk, however, to give Joey the feeling that he was trick-or-treating by himself.

After the traditional query, "Trick or treat?" I asked Joey to hold open his right hand. I placed three Hershey's Kisses in his palm and asked him to hold them there for a moment. "You can also get one of these two Snickers bars," I said, showing him a small one and a large one. "In fact, if you give me one of those Hershey's Kisses I will give you this smaller Snickers bar. And if you give me two of your Hershey's Kisses, I will give you this larger Snickers bar."

Now a kid may dress up like a giant spider, but that doesn't mean he's stupid. The small Snickers bar weighed one ounce, and the large Snickers bar weighed two ounces. All Joey had to do was give me one additional Hershey's Kiss (about 0.16 ounce) and he would get an extra ounce of Snickers. This deal might have stumped a rocket scientist, but for a nine-year-old boy, the computation was easy: he'd get more than six times the return on investment (in the net weight of chocolate) if he went for the larger Snickers bar. In a flash Joey put two of his Kisses into my hand, took the two-ounce Snickers bar, and dropped it into his bag.

Joey wasn't alone in making this snap decision. All but

one of the kids to whom I presented this offer traded in two Kisses for the bigger candy bars.

Zoe was the next kid to walk down the street. She was dressed as a princess, in a long white dress, with a magic wand in one hand and an orange Halloween pumpkin bucket in the other. Her younger sister was resting comfortably in their father's arms, looking cute and cuddly in her bunny outfit. As they approached, Zoe called out, in a high, cute voice, "Trick or treat!" In the past I admit that I have sometimes devilishly replied, "Trick!" Most kids stand there, baffled, having never thought through their question to see that it allowed an alternative answer.

In this case I gave Zoe her treat—three Hershey's Kisses. But I did have a trick up my sleeve. I offered little Zoe a deal: a choice between getting a large Snickers bar in exchange for one of her Hershey's Kisses, or getting the small Snickers bar for FREE! without giving up any Hershey's Kisses.

Now, a bit of rational calculation (which in Joey's case was amply demonstrated) would show that the best deal is to forgo the free small Snickers bar, pay the cost of one additional Hershey's Kiss, and go for the large Snickers bar. On an ounce-for-ounce comparison, it was far better to give up one additional Hershey's Kiss and get the larger Snickers bar (two ounces) instead of a smaller Snickers bar (one ounce). This logic was perfectly clear to Joe and the kids who encountered the condition in which both Snickers bars had a cost. But what would Zoe do? Would her clever kid's mind make that rational choice—or would the fact that the small Snickers bar was FREE! blind her to the rationally correct answer?

As you might have guessed by now, Zoe, and the other kids to whom I offered the same deal, was completely blinded by FREE! About 70 percent of them gave up the better deal, and took the worse deal just because it was FREE!

Just in case you think Kristina, Nina, and I make a habit of picking on kids, I'll mention that we repeated the experiment with bigger kids, in fact students at the MIT student center. The results replicated the pattern we saw on Halloween. Indeed, the draw of zero cost is not limited to monetary transactions. Whether it's products or money, we just can't resist the gravitational pull of FREE!

So DO YOU think you have a handle on FREE!?

OK. Here's a quiz. Suppose I offered you a choice between a free $10 Amazon gift certificate and a $20 gift certificate for seven dollars. Think quickly. Which would you take?

If you jumped for the FREE! certificate, you would have been like most of the people we tested at one of the malls in Boston. But look again: a $20 gift certificate for seven dollars delivers a $13 profit. That's clearly better than getting a $10 certificate free (earning $10). Can you see the irrational behavior in action?*

LET ME TELL you a story that describes the real influence of FREE! on our behavior. A few years ago, Amazon.com started offering free shipping of orders over a certain amount. Someone who purchased a single book for $16.95 might pay an additional $3.95 for shipping, for instance. But if the cus-

---

*Similar to the other experiments, when we increased the cost of both certificates by $1, making the $10 certificate cost $1 and the $20 certificate cost $8, the majority jumped for the $20 certificate.

tomer bought another book, for a total of $31.90, they would get their shipping FREE!

Some of the purchasers probably didn't want the second book (and I am talking here from personal experience) but the FREE! shipping was so tempting that to get it, they were willing to pay the cost of the extra book. The people at Amazon were very happy with this offer, but they noticed that in one place—France—there was no increase in sales. Is the French consumer more rational than the rest of us? Unlikely. Rather, it turned out, the French customers were reacting to a different deal.

Here's what happened. Instead of offering FREE! shipping on orders over a certain amount, the French division priced the shipping for those orders at one franc. Just one franc—about 20 cents. This doesn't seem very different from FREE! but it was. In fact, when Amazon changed the promotion in France to include free shipping, France joined all the other countries in a dramatic sales increase. In other words, whereas shipping for one franc—a real bargain—was virtually ignored by the French, FREE! shipping caused an enthusiastic response.

America Online (AOL) had a similar experience several years ago when it switched from pay-per-hour service to a monthly payment schedule (in which you could log in as many hours as you wanted for a fixed $19.95 per month). In preparation for the new price structure, AOL geared up for what it estimated would be a small increase in demand. What did it get? An overnight increase from 140,000 to 236,000 customers logging into the system, and a doubling of the average time online. That may seem good—but it wasn't good. AOL's customers encountered busy phone lines, and soon AOL was forced to lease services from other online providers (who were only too happy to sell bandwidth to AOL—at the premium of snow shovels in a snowstorm). What Bob Pittman (the president of AOL at the time) didn't realize was that consumers would respond to the allure of FREE! like starving people at a buffet.

WHEN CHOOSING BETWEEN two products, then, we often overreact to the free one. We might opt for a FREE! checking account (with no benefits attached) rather than one that costs five dollars a month. But if the five-dollar checking account includes free traveler's checks, online billing, etc., and the FREE! one doesn't, we may end up spending more for this package of services with the FREE! account than with the five-dollar account. Similarly, we might choose a mortgage with no closing costs, but with interest rates and fees that are off the wall; and we might get a product we don't really want simply because it comes with a free gift.

My most recent personal encounter with this involved a car. When I was looking for a new car a few years ago, I knew that I really should buy a minivan. In fact, I had read up on Honda minivans and knew all about them. But then an Audi caught my eye, at first through an appealing offer—FREE! oil changes for the next three years. How could I resist?

To be perfectly honest, the Audi was sporty and red, and I was still resisting the idea of being a mature and responsible father to two young kids. It wasn't as if the free oil change completely swayed me, but its influence on me was, from a rational perspective, unjustifiably large. Just because it was FREE! it served as an additional allure that I could cling to.

So I bought the Audi—and the FREE! oil. (A few months later, while I was driving on a highway, the transmission broke—but that is a different story.) Of course, with a cooler head I might have made a more rational calculation. I drive about 7,000 miles a year; the oil needs to be changed every 10,000 miles; and the cost per change is about $75. Over three years, then, I would save about $150, or about 0.5 percent of the purchase price of the car—not a good reason to base my decision on. It gets worse, though: now I have an Audi that is packed to the ceiling with action figures, a stroller, a bike, and other kids' paraphernalia. Oh, for a minivan.

THE CONCEPT OF zero also applies to time. Time spent on one activity, after all, is time taken away from another. So if we spend 45 minutes in a line waiting for our turn to get a FREE! taste of ice cream, or if we spend half an hour filling out a long form for a tiny rebate, there is something else that we are not doing with our time.

My favorite personal example is free-entrance day at a museum. Despite the fact that most museums are not very expensive, I find it much more appealing to satisfy my desire for art when the price is zero. Of course I am not alone in this desire. So on these days I usually find that the museum is overcrowded, the line is long, it is hard to see anything, and fighting the crowds around the museum and in the cafeteria is unpleasant. Do I realize that it is a mistake to go to a museum when it is free? You bet I do—but I go nevertheless.

ZERO MAY ALSO affect food purchases. Food manufacturers have to convey all kinds of information on the side of the box. They have to tell us about the calories, fat content, fiber, etc. Is it possible that the same attraction we have to zero price could also apply to zero calories, zero trans fats, zero carbs, etc.? If the same general rules apply, Pepsi will sell more cans if the label says "zero calories" than if it says "one calorie."

Suppose you are at a bar, enjoying a conversation with some friends. With one brand you get a calorie-free beer, and with another you get a three-calorie beer. Which brand will make you feel that you are drinking a really light beer? Even though the difference between the two beers is negligible, the zero-calorie beer will increase the feeling that you're doing the right thing, healthwise. You might even feel so good that you go ahead and order a plate of fries.

So YOU CAN maintain the status quo with a 20-cent fee (as in the case of Amazon's shipping in France), or you can start a stampede by offering something FREE! Think how powerful that idea is! Zero is not just another discount. Zero is a different place. The difference between two cents and one cent is small. But the difference between one cent and zero is huge!

If you are in business, and understand that, you can do some marvelous things. Want to draw a crowd? Make something FREE! Want to sell more products? Make part of the purchase FREE!

Similarly, we can use FREE! to drive social policy. Want people to drive electric cars? Don't just lower the registration and inspection fees—eliminate them, so that you have created FREE! In the same way, if health is your concern, focus on early detection as a way to eliminate the progression of severe illnesses. Want people to do the right thing—in terms of getting regular colonoscopies, mammograms, cholesterol checks, diabetes checks, and such? Don't just decrease the cost (by decreasing the co-pay). Make these critical procedures FREE!

I don't think most policy strategists realize that FREE! is an ace in their hand, let alone know how to play it. It's certainly counterintuitive, in these times of budget cutbacks, to make something FREE! But when we stop to think about it, FREE! can have a great deal of power, and it makes a lot of sense.

APPENDIX: CHAPTER 3

Let me explain how the logic of standard economic theory would apply to our setting. When a person can select one and only one of two chocolates, he needs to consider not the absolute value of each chocolate but its relative value—what he gets and what he gives up. As a first step the rational consumer needs to compute the relative net benefits of the two chocolates (the value of the expected taste minus the cost), and make a decision based on which chocolate has the larger net benefit. How would this look when the cost of the Lindt truffle was 15 cents and the cost of the Hershey's Kiss was one cent? The rational consumer would estimate the amount of pleasure he expects to get from the truffle and the Kiss (let's say this is 50 pleasure units and five pleasure units, respectively) and subtract the displeasure he would get from paying 15 cents and one cent (let's say this is 15 displeasure units and one displeasure unit, respectively). This would give him a total expected pleasure of 35 pleasure units (50 − 15) for the truffle, and a total expected pleasure of four pleasure units (5 − 1) for the Kiss. The truffle leads by 31 points, so it's an easy choice—the truffle wins hands down.

What about the case when the cost is reduced by the same amount for both products? (Truffles cost 14 cents and the Kiss is free.) The same logic applies. The taste of the chocolates has not changed, so the rational consumer would estimate the pleasure to be 50 and five pleasure units, respectively. What has changed is the displeasure. In this setting the rational consumer would have a lower level of displeasure for both chocolates because the prices have been reduced by one cent (and one displeasure unit). Here is the main point: because both products were discounted by the same amount, their relative difference would be unchanged. The total expected pleasure for the truffle would now be 36 pleasure units (50 − 14), and the total expected pleasure for the Kiss would now be five pleasure units (5 − 0). The truffle leads by the same 31 points, so it should be the same easy choice. The truffle wins hands down.

This is how the pattern of choice *should* look, if the only forces at play were those of a rational cost-benefit analysis. The fact that the results from our experiments are so different tells us loud and clear that something else is going on, and that the price of zero plays a unique role in our decisions.

CHAPTER 4

The Cost of Social Norms

Why We Are Happy to Do Things, but Not When We Are Paid to Do Them

You are at your mother-in-law's house for Thanksgiving dinner, and what a sumptuous spread she has put on the table for you! The turkey is roasted to a golden brown; the stuffing is homemade and exactly the way you like it. Your kids are delighted: the sweet potatoes are crowned with marshmallows. And your wife is flattered: her favorite recipe for pumpkin pie has been chosen for dessert.

The festivities continue into the late afternoon. You loosen your belt and sip a glass of wine. Gazing fondly across the table at your mother-in-law, you rise to your feet and pull out your wallet. "Mom, for all the love you've put into this, how much do I owe you?" you say sincerely. As silence descends on the gathering, you wave a handful of bills. "Do you think three hundred dollars will do it? No, wait, I should give you four hundred!"

This is not a picture that Norman Rockwell would have painted. A glass of wine falls over; your mother-in-law stands up red-faced; your sister-in-law shoots you an angry look; and your niece bursts into tears. Next year's Thanksgiving celebration, it seems, may be a frozen dinner in front of the television set.

WHAT'S GOING ON here? Why does an offer for direct payment put such a damper on the party? As Margaret Clark, Judson Mills, and Alan Fiske suggested a long time ago, the answer is that we live simultaneously in two different worlds—one where social norms prevail, and the other where market norms make the rules. The social norms include the friendly requests that people make of one another. Could you help me move this couch? Could you help me change this tire? Social norms are wrapped up in our social nature and our need for community. They are usually warm and fuzzy. Instant paybacks are not required: you may help move your neighbor's couch, but this doesn't mean he has to come right over and move yours. It's like opening a door for someone: it provides pleasure for both of you, and reciprocity is not immediately required.

The second world, the one governed by market norms, is very different. There's nothing warm and fuzzy about it. The exchanges are sharp-edged: wages, prices, rents, interest, and costs-and-benefits. Such market relationships are not necessarily evil or mean—in fact, they also include self-reliance, inventiveness, and individualism—but they do imply comparable benefits and prompt payments. When you are in the domain of market norms, you get what you pay for—that's just the way it is.

When we keep social norms and market norms on their separate paths, life hums along pretty well. Take sex, for instance. We may have it free in the social context, where it is, we hope, warm and emotionally nourishing. But there's also market sex, sex that is on demand and that costs money. This seems pretty straightforward. We don't have husbands (or wives) coming home asking for a $50 trick; nor do we have prostitutes hoping for everlasting love.

When social and market norms collide, trouble sets in. Take sex again. A guy takes a girl out for dinner and a movie, and he pays the bills. They go out again, and he pays the bills once more. They go out a third time, and he's still springing for the meal and the entertainment. At this point, he's hoping for at least a passionate kiss at the front door. His wallet is getting perilously thin, but worse is what's going on in his head: he's having trouble reconciling the social norm (courtship) with the market norm (money for sex). On the fourth date he casually mentions how much this romance is costing him. Now he's crossed the line. Violation! She calls him a beast and storms off. He should have known that one can't mix social and market norms—especially in this case—without implying that the lady is a tramp. He should also have remembered the immortal words of Woody Allen: "The most expensive sex is free sex."

A FEW YEARS ago, James Heyman (a professor at the University of St. Thomas) and I decided to explore the effects of social and market norms. Simulating the Thanksgiving incident would have been wonderful, but considering the damage we might have done to our participants' family relationships, we chose something more mundane. In fact, it was one of the most boring tasks we could find (there is a tradition in social science of using very boring tasks).

In this experiment, a circle was presented on the left side of a computer screen and a box was presented on the right. The task was to drag the circle, using the computer mouse, onto the square. Once the circle was successfully dragged to the square, it disappeared from the screen and a new circle appeared at the starting point. We asked the participants to drag as many circles as they could, and we measured how many circles they dragged within five minutes. This was our measure of their labor output—the effort that they would put into this task.

How could this setup shed light on social and market exchanges? Some of the participants received five dollars for participating in the short experiment. They were given the money as they walked into the lab; and they were told that at the end of the five minutes, the computer would alert them that the task was done, at which point they were to leave the lab. Because we paid them for their efforts, we expected them to apply market norms to this situation and act accordingly.

Participants in a second group were presented with the same basic instructions and task; but for them the reward was much lower (50 cents in one experiment and 10 cents in the other). Again we expected the participants to apply market norms to this situation and act accordingly.

Finally, we had a third group, to whom we introduced the tasks as a social request. We didn't offer the participants in this group anything concrete in return for their effort; nor did we mention money. It was merely a favor that we asked of them. We expected these participants to apply social norms to the situation and act accordingly.

How hard did the different groups work? In line with the ethos of market norms, those who received five dollars dragged on average 159 circles, and those who received 50 cents dragged on average 101 circles. As expected, more money caused our participants to be more motivated and work harder (by about 50 percent).

What about the condition with no money? Did these participants work less than the ones who got the low monetary payment—or, in the absence of money, did they apply social norms to the situation and work harder? The results showed that on average they dragged 168 circles, much more than those who were paid 50 cents, and just slightly more than those who were paid five dollars. In other words, our participants worked harder under the nonmonetary social norms than for the almighty buck (OK, 50 cents).

Perhaps we should have anticipated this. There are many examples to show that people will work more for a cause than for cash. A few years ago, for instance, the AARP asked some lawyers if they would offer less expensive services to needy retirees, at something like $30 an hour. The lawyers said no. Then the program manager from AARP had a brilliant idea: he asked the lawyers if they would offer free services to needy retirees. Overwhelmingly, the lawyers said yes.

What was going on here? How could zero dollars be more attractive than $30? When money was mentioned, the lawyers used market norms and found the offer lacking, relative to their market salary. When no money was mentioned they used social norms and were willing to volunteer their time. Why didn't they just accept the $30, thinking of themselves as volunteers who received $30? Because once market norms enter our considerations, the social norms depart.

A similar lesson was learned by Nachum Sicherman, an economics professor at Columbia, who was taking martial arts lessons in Japan. The sensei (the master teacher) was not charging the group for the training. The students, feeling that this was unfair, approached the master one day and suggested that they pay him for his time and effort. Setting down his bamboo *shinai*, the master calmly replied that if he charged them, they would not be able to afford him.

IN THE PREVIOUS experiment, then, those who got paid 50 cents didn't say to themselves, "Good for me; I get to do this favor for these researchers, and I am getting some money out of this," and continue to work harder than those who were paid nothing. Instead they switched themselves over to the market norms, decided that 50 cents wasn't much, and worked halfheartedly. In other words, when the market norms entered the lab, the social norms were pushed out.

But what would happen if we replaced the payments with a gift? Surely your mother-in-law would accept a good bottle of wine at dinner. Or how about a housewarming present (such as an eco-friendly plant) for a friend? Are gifts methods of exchange that keep us within the social exchange norms? Would participants receiving such gifts switch out of the social norms and into market norms, or would offering gifts as rewards maintain the participants in the social world?

To find out just where gifts fall on the line between social and market norms, James and I decided on a new experiment. This time, we didn't offer our participants money for dragging circles across a computer screen; we offered them gifts instead. We replaced the 50-cent reward with a Snickers bar (worth about 50 cents), and the five-dollar incentive with a box of Godiva chocolates (worth about five dollars).

The participants came to the lab, got their reward, worked as much as they liked, and left. Then we looked at the results. As it turned out, all three experimental groups worked about equally hard during the task, regardless of whether they got a small Snickers bar (these participants dragged on average 162 circles), the Godiva chocolates (these participants dragged on average 169 circles), or nothing at all (these participants dragged on average 168 circles). The conclusion: no one is offended by a small gift, because even small gifts keep us in the social exchange world and away from market norms.

BUT WHAT WOULD HAPPEN if we mixed the signals for the two types of norms? What would happen if we blended the market norm with the social norm? In other words, if we said that we would give them a "50-*cent* Snickers bar" or a "*five-dollar* box of Godiva chocolates," what would the participants do? Would a "50-cent Snickers bar" make our participants work as hard as a "Snickers bar" made them work; or would it make them work halfheartedly, as the 50-cents made them work? Or would it be somewhere in the middle? The next experiment tested these ideas.

As it turned out, the participants were not motivated to work at all when they got the 50-cent Snickers bar, and in fact the effort they invested was the same as when they got a payment of 50 cents. They reacted to the explicitly priced gift in exactly the way they reacted to cash, and the gift no longer invoked social norms—by the mention of its cost, the gift had passed into the realm of market norms.

By the way, we replicated the setup later when we asked passersby whether they would help us unload a sofa from a truck. We found the same results. People are willing to work free, and they are willing to work for a reasonable wage; but offer them just a small payment and they will walk away. Gifts are also effective for sofas, and offering people a gift, even a small one, is sufficient to get them to help; but mention what the gift cost you, and you will see the back of them faster than you can say market norms.

THESE RESULTS SHOW that for market norms to emerge, it is sufficient to mention money (even when no money changes hands). But, of course, market norms are not just about effort—they relate to a broad range of behaviors, including self-reliance, helping, and individualism. Would simply getting people to think about money influence them to behave differently in these respects? This premise was explored in a set of fantastic experiments by Kathleen Vohs (a professor at the University of Minnesota), Nicole Mead (a graduate student at Florida State University), and Miranda Goode (a graduate student at the University of British Columbia).

They asked the participants in their experiments to complete a "scrambled-sentence task," that is, to rearrange sets of words to form sentences. For the participants in one group, the task was based on neutral sentences (for example, "It's cold outside"); for the other group, the task was based on sentences or phrases related to money (for example, "High-paying salary"*). Would thinking about money in this manner be sufficient to change the way participants behave?

In one of the experiments, the participants finished the unscrambling task and were then given a difficult puzzle, in which they had to arrange 12 disks into a square. As the experimenter left the room, he told them that they could come to him if they needed any help. Who do you think asked for

---

*This general procedure is called priming, and the unscrambling task is used to get participants to think about a particular topic—without direct instructions to do so.

help sooner—those who had worked on the "salary" sentences, with their implicit suggestion of money; or those who had worked on the "neutral" sentences, about the weather and other such topics? As it turned out, the students who had first worked on the "salary" task struggled with the puzzle for about five and a half minutes before asking for help, whereas those who had first worked on the neutral task asked for help after about three minutes. Thinking about money, then, made the participants in the "salary" group more self-reliant and less willing to ask for help.

But these participants were also less willing to help others. In fact, after thinking about money these participants were less willing to help an experimenter enter data, less likely to assist another participant who seemed confused, and less likely to help a "stranger" (an experimenter in disguise) who "accidentally" spilled a box of pencils.

Overall, the participants in the "salary" group showed many of the characteristics of the market: they were more selfish and self-reliant; they wanted to spend more time alone; they were more likely to select tasks that required individual input rather than teamwork; and when they were deciding where they wanted to sit, they chose seats farther away from whomever they were told to work with. Indeed, just thinking about money makes us behave as most economists believe we behave—and less like the social animals we are in our daily lives.

This leads me to a final thought: when you're in a restaurant with a date, for heaven's sake don't mention the price of the selections. Yes, they're printed clearly on the menu. Yes, this might be an opportunity to impress your date with the caliber of the restaurant. But if you rub it in, you'll be likely to shift your relationship from the social to the market norm.

Yes, your date may fail to recognize how much this meal is setting you back. Yes, your mother-in-law may assume that the bottle of wine you've presented is a $10 blend, when it's a $60 special reserve merlot. That's the price you have to pay, though, to keep your relationships in the social domain and away from market norms.

So we live in two worlds: one characterized by social exchanges and the other characterized by market exchanges. And we apply different norms to these two kinds of relationships. Moreover, introducing market norms into social exchanges, as we have seen, violates the social norms and hurts the relationships. Once this type of mistake has been committed, recovering a social relationship is difficult. Once you've offered to pay for the delightful Thanksgiving dinner, your mother-in-law will remember the incident for years to come. And if you've ever offered a potential romantic partner the chance to cut to the chase, split the cost of the courting process, and simply go to bed, the odds are that you will have wrecked the romance forever.

My good friends Uri Gneezy (a professor at the University of California at San Diego) and Aldo Rustichini (a professor at the University of Minnesota) provided a very clever test of the long-term effects of a switch from social to market norms.

A few years ago, they studied a day care center in Israel to determine whether imposing a fine on parents who arrived late to pick up their children was a useful deterrent. Uri and Aldo concluded that the fine didn't work well, and in fact it had long-term negative effects. Why? Before the fine was introduced, the teachers and parents had a social contract, with social norms about being late. Thus, if parents were late—as they occasionally were—they felt guilty about it—and their guilt compelled them to be more prompt in picking up their kids in the future. (In Israel, guilt seems to be an effective way to get compliance.) But once the fine was imposed, the day care center had inadvertently replaced the social norms with market norms. Now that the parents were *paying* for their tardiness, they interpreted the situation in terms of market norms. In other words, since they were being fined, they could decide for themselves whether to be late or not, and they frequently chose to be late. Needless to say, this was not what the day care center intended.

BUT THE REAL story only started here. The most interesting part occurred a few weeks later, when the day care center removed the fine. Now the center was back to the social norm. Would the parents also return to the social norm? Would their guilt return as well? Not at all. Once the fine was removed, the behavior of the parents didn't change. They continued to pick up their kids late. In fact, when the fine was removed, there was a slight increase in the number of tardy pickups (after all, both the social norms and the fine had been removed).

This experiment illustrates an unfortunate fact: when a social norm collides with a market norm, the social norm goes away for a long time. In other words, social relationships are not easy to reestablish. Once the bloom is off the rose—once a social norm is trumped by a market norm—it will rarely return.

THE FACT THAT we live in both the social world and the market world has many implications for our personal lives. From time to time, we all need someone to help us move something, or to watch our kids for a few hours, or to take in our mail when we're out of town. What's the best way to motivate our friends and neighbors to help us? Would cash do it—a gift, perhaps? How much? Or nothing at all? This social dance, as I'm sure you know, isn't easy to figure out—especially when there's a risk of pushing a relationship into the realm of a market exchange.

Here are some answers. Asking a friend to help move a large piece of furniture or a few boxes is fine. But asking a friend to help move a lot of boxes or furniture is not—especially if the friend is working side by side with movers who are getting paid for the same task. In this case, your friend might begin to feel that he's being used. Similarly, asking your neighbor (who happens to be a lawyer) to bring in your mail while you're on vacation is fine. But asking him to spend the same amount of time preparing a rental contract for you—free—is not.

THE DELICATE BALANCE between social and market norms is also evident in the business world. In the last few decades companies have tried to market themselves as social companions—that is, they'd like us to think that they and we are family, or at least are friends who live on the same cul-de-sac. "Like a good neighbor, State Farm is there" is one familiar slogan. Another is Home Depot's gentle urging: "You can do it. We can help."

Whoever started the movement to treat customers socially had a great idea. If customers and a company are family, then the company gets several benefits. Loyalty is paramount. Minor infractions—screwing up your bill and even imposing a modest hike in your insurance rates—are accommodated.

Relationships of course have ups and downs, but overall they're a pretty good thing.

But here's what I find strange: although companies have poured billions of dollars into marketing and advertising to create social relationships—or at least an impression of social relationships—they don't seem to understand the nature of a social relationship, and in particular its risks.

For example, what happens when a customer's check bounces? If the relationship is based on market norms, the bank charges a fee, and the customer shakes it off. Business is business. While the fee is annoying, it's nonetheless acceptable. In a social relationship, however, a hefty late fee—rather than a friendly call from the manager or an automatic fee waiver—is not only a relationship-killer; it's a stab in the back. Consumers will take personal offense. They'll leave the bank angry and spend hours complaining to their friends about this awful bank. After all, this was a relationship framed as a social exchange. No matter how many cookies, slogans, and tokens of friendship a bank provides, one violation of the social exchange means that the consumer is back to the market exchange. It can happen that quickly.

What's the upshot? If you're a company, my advice is to remember that you can't have it both ways. You can't treat your customers like family one moment and then treat them impersonally—or, even worse, as a nuisance or a competitor— a moment later when this becomes more convenient or profitable. This is not how social relationships work. If you want a social relationship, go for it, but remember that you have to maintain it under all circumstances.

On the other hand, if you think you may have to play tough from time to time—charging extra for additional services or rapping knuckles swiftly to keep the consumers in line—you might not want to waste money in the first place on making your company the fuzzy feel-good choice. In that case, stick to a simple value proposition: state what you give and what you expect in return. Since you're not setting up any social norms or expectations, you also can't violate any—after all, it's just business.

COMPANIES HAVE ALSO tried to establish social norms with their employees. It wasn't always this way. Years ago, the workforce of America was more of an industrial, market-driven exchange. Back then it was often a nine-to-five, time-clock kind of mentality. You put in your 40 hours and you got your paycheck on Friday. Since workers were paid by the hour, they knew exactly when they were working for the man, and when they weren't. The factory whistle blew (or the corporate equivalent took place), and the transaction was finished. This was a clear market exchange, and it worked adequately for both sides.

Today companies see an advantage in creating a social exchange. After all, in today's market we're the makers of intangibles. Creativity counts more than industrial machines. The partition between work and leisure has likewise blurred. The people who run the workplace want us to think about work while we're driving home and while we're in the shower. They've given us laptops, cell phones, and BlackBerries to bridge the gap between the workplace and home.

Further blurring the nine-to-five workday is the trend in many companies to move away from hourly rates to monthly pay. In this 24/7 work environment social norms have a great advantage: they tend to make employees passionate, hard-working, flexible, and concerned. In a market where employees' loyalty to their employers is often wilting, social norms are one of the best ways to make workers loyal, as well as motivated.

Open-source software shows the potential of social norms. In the case of Linux and other collaborative projects, you can post a problem about a bug on one of the bulletin boards and see how fast someone, or often many people, will react to your request and fix the software—using their own leisure time. Could you pay for this level of service? Most likely. But if you had to hire people of the same caliber they would cost you an arm and a leg. Rather, people in these communities are happy to give their time to society at large (for which they get the same social benefits we all get from helping a friend paint a room). What can we learn from this that is applicable to the business world? There are social rewards that strongly motivate behavior—and one of the least used in corporate life is the encouragement of social rewards and reputation.

IN TREATING THEIR EMPLOYEES—much as in treating their customers—companies must understand their implied long-term commitment. If employees promise to work harder to achieve an important deadline (even canceling family obligations for it), if they are asked to get on an airplane at a moment's notice to attend a meeting, then they must get something similar in return—something like support when they are sick, or a chance to hold on to their jobs when the market threatens to take their jobs away.

Although some companies have been successful in creating social norms with their workers, the current obsession with short-term profits, outsourcing, and draconian cost cutting threatens to undermine it all. In a social exchange, after all, people believe that if something goes awry the other party will be there for them, to protect and help them. These beliefs are not spelled out in a contract, but they are general obligations to provide care and help in times of need.

Again, companies cannot have it both ways. In particular, I am worried that the recent cuts we see in employees' benefits—child care, pensions, flextime, exercise rooms, the cafeteria, family picnics, etc.—are likely to come at the expense of the social exchange and thus affect workers' productivity. I am particularly worried that cuts and changes in medical benefits are likely to transform much of the employer-employee social relationship to a market relationship.

If companies want to benefit from the advantages of social norms, they need to do a better job of cultivating those norms. Medical benefits, and in particular comprehensive medical coverage, are among the best ways a company can express its side of the social exchange. But what are many companies doing? They are demanding high deductibles in their insurance plans, and at the same time are reducing the scope of benefits. Simply put, they are undermining the social contract between the company and the employees and replacing it with market norms. As companies tilt the board, and employees slide from social norms to the realm of market norms, can we blame them for jumping ship when a better offer appears? It's really no surprise that "corporate loyalty," in terms of the loyalty of employees to their companies, has become an oxymoron.

Organizations can also think consciously about how people react to social and market norms. Should you give an employee a gift worth $1,000 or pay him or her an extra $1,000 in cash? Which is better? If you ask the employees, the majority will most likely prefer cash over the gift. But the gift has its value, though this is sometimes ill understood—it can provide a boost to the social relationship between the employer and the employee, and by doing so provide long-term benefits to everyone. Think of it this way: who do you suppose is likely to work harder, show more loyalty, and truly love his work more—someone who is getting $1,000 in cash or someone who is getting a personal gift?

Of course, a gift is a symbolic gesture. And to be sure, no one is going to work for gifts rather than a salary. For that matter, no one is going to work for nothing. But if you look at companies like Google, which offers a wide variety of benefits for employees (including free gourmet lunches), you can see how much goodwill is created by emphasizing the social side of the company-worker relationship. It's remarkable how much work companies (particularly start-ups) can get out of people when social norms (such as the excitement of building something together) are stronger than market norms (such as salaries stepping up with each promotion).

If corporations started thinking in terms of social norms, they would realize that these norms build loyalty and—more important—make people want to extend themselves to the degree that corporations need today: to be flexible, concerned, and willing to pitch in. That's what a social relationship delivers.

THIS QUESTION OF social norms in the workplace is one we should be thinking about frequently. America's productivity depends increasingly on the talent and efforts of its workers. Could it be that we are driving business from the realm of social norms into market norms? Are workers thinking in terms of money, rather than the social values of loyalty and trust? What will that do to American productivity in the long run, in terms of creativity and commitment? And what of the "social contract" between government and the citizen? Is that at risk as well?

At some level we all know the answers. We understand, for instance, that a salary alone will not motivate people to risk their lives. Police officers, firefighters, soldiers—they don't die for their weekly pay. It's the social norms—pride in their profession and a sense of duty—that will motivate them to give up their lives and health. A friend of mine in Miami once accompanied a U.S. customs agent on a patrol of the offshore waters. The agent carried an assault rifle and could certainly have pounded several holes into a fleeing drug boat. But had he ever done so? No way, he replied. He wasn't about to get himself killed for the government salary he received. In fact, he confided, his group had an unspoken agreement with the drug couriers: the feds wouldn't fire if the drug dealers didn't fire. Perhaps that's why we rarely (if ever) hear about gun battles on the edges of America's "war on drugs."

How could we change this situation? First, we could make the federal salary so good that the customs agent would be willing to risk his life for it. But how much money is that? Compensation equal to what the typical drug trafficker gets for racing a boat from the Bahamas to Miami? Alternatively, we could elevate the social norm, making the officer feel that his mission is worth more than his base pay—that we honor him (as we honor our police and firefighters) for a job which not only stabilizes the structure of society but also saves our kids from all kinds of dangers. That would take some inspirational leadership, of course, but it could be done.

Let me describe how that same thought applies to the world of education. I recently joined a federal committee on incentives and accountability in public education. This is one aspect of social and market norms that I would like to explore in the years to come. Our task is to reexamine the "No Child Left Behind" policy, and to help find ways to motivate students, teachers, administrators, and parents.

My feeling so far is that standardized testing and performance-based salaries are likely to push education from social norms to market norms. The United States already spends more money per student than any other Western society. Would it be wise to add more money? The same consideration applies to testing: we are already testing very frequently, and more testing is unlikely to improve the quality of education.

I suspect that one answer lies in the realm of social norms. As we learned in our experiments, cash will take you only so far—social norms are the forces that can make a difference in the long run. Instead of focusing the attention of the teachers, parents, and kids on test scores, salaries, and competition, it might be better to instill in all of us a sense of purpose, mission, and pride in education. To do this we certainly can't take the path of market norms. The Beatles proclaimed some time ago that you "Can't Buy Me Love" and this also applies to the love of learning—you can't buy it; and if you try, you might chase it away.

So how can we improve the educational system? We should probably first rethink school curricula, and link them in more obvious ways to social goals (elimination of poverty and crime, elevation of human rights, etc.), technological goals (boosting energy conservation, space exploration, nanotechnology, etc.), and medical goals (cures for cancer, diabetes, obesity, etc.) that we care about as a society. This way the students, teachers, and parents might see the larger point in education and become more enthusiastic and motivated about it. We should also work hard on making education a goal in itself, and stop confusing the number of hours students spend in school with the quality of the education they get. Kids can get excited about many things (baseball, for example), and it is our challenge as a society to make them want to know as much about Nobel laureates as they now know about baseball players. I am not suggesting that igniting a social passion for education is simple; but if we succeed in doing so, the value could be immense.

MONEY, AS IT turns out, is very often the most expensive way to motivate people. Social norms are not only cheaper, but often more effective as well.

So what good is money? In ancient times, money made trading easier: you didn't have to sling a goose over your back when you went to market, or decide what section of the goose was equivalent to a head of lettuce. In modern times money has even more benefits, as it allows us to specialize, borrow, and save.

But money has also taken on a life of its own. As we have seen, it can remove the best in human interactions. So do we need money? Of course we do. But could there be some aspects of our life that would be, in some ways, better without it?

That's a radical idea, and not an easy one to imagine. But a few years ago I had a taste of it. At that time, I got a phone call from John Perry Barlow, a former lyricist for the Grateful Dead, inviting me to an event that proved to be both an important personal experience and an interesting exercise in creating a moneyless society. Barlow told me that I had to come to Burning Man with him, and that if I did, I would feel as if I had come home. Burning Man is an annual week-long event of self-expression and self-reliance held in Black Rock Desert, Nevada, regularly attended by more than 40,000 people. Burning Man started in 1986 on Baker Beach in San Francisco, when a small crowd designed, built, and eventually set fire to an eight-foot wooden statue of a man and a smaller wooden dog. Since then the size of the man being burned and the number of people who attend the festivities has grown considerably, and the event is now one of the largest art festivals, and an ongoing experiment in temporary community.

Burning Man has many extraordinary aspects, but for me one of the most remarkable is its rejection of market norms. Money is not accepted at Burning Man. Rather, the whole place works as a gift exchange economy—you give things to other people, with the understanding that they will give something back to you (or to someone else) at some point in the future. Thus, people who can cook might fix a meal. Psychologists offer free counseling sessions. Masseuses massage those lying on tables before them. Those who have water offer showers. People give away drinks, homemade jewelry, and hugs. (I made some puzzles at the hobby shop at MIT, and gave them to people. Mostly, people enjoyed trying to solve them.)

At first this was all very strange, but before long I found myself adopting the norms of Burning Man. I was surprised, in fact, to find that Burning Man was the most accepting, social, and caring place I had ever been. I'm not sure I could easily survive in Burning Man for all 52 weeks of the year. But this experience has convinced me that life with fewer market norms and more social norms would be more satisfying, creative, fulfilling, and fun.

The answer, I believe, is not to re-create society as Burning Man, but to remember that social norms can play a far greater role in society than we have been giving them credit for. If we contemplate how market norms have gradually taken over our lives in the past few decades—with their emphasis on higher salaries, more income, and more spending— we may recognize that a return to some of the old social norms might not be so bad after all. In fact, it might bring quite a bit of the old civility back to our lives.

CHAPTER 5

The Influence of Arousal

Why Hot Is Much Hotter Than We Realize

A sk most twentysomething male college students whether they would ever attempt unprotected sex and they will quickly recite chapter and verse about the risk of dreaded diseases and pregnancy. Ask them in any dispassionate circumstances—while they are doing homework or listening to a lecture—whether they'd enjoy being spanked, or enjoy sex in a threesome with another man, and they'll wince. No way, they'd tell you. Furthermore, they'd narrow their eyes at you and think, What kind of sicko are you anyhow, asking these questions in the first place?

In 2001, while I was visiting Berkeley for the year, my friend, academic hero, and longtime collaborator George Loewenstein and I invited a few bright students to help us understand the degree to which rational, intelligent people can predict how their attitudes will change when they are in an impassioned state. In order to make this study realistic, we needed to measure the participants' responses while they were smack in the midst of such an emotional state. We could have made our participants feel angry or hungry, frustrated or annoyed. But we preferred to have them experience a pleasurable emotion.

We chose to study decision making under sexual arousal—not because we had kinky predilections ourselves, but because understanding the impact of arousal on behavior might help society grapple with some of its most difficult problems, such as teen pregnancy and the spread of HIV-AIDS. There are sexual motivations everywhere we look, and yet we understand very little about how these influence our decision making.

Moreover, since we wanted to understand whether participants would be able to predict how they would behave in a particular emotional state, the emotion needed to be one that was already quite familiar to them. That made our decision easy. If there's anything predictable and familiar about twentysomething male college students, it's the regularity with which they experience sexual arousal.

ROY, AN AFFABLE, studious biology major at Berkeley, is in a sweat—and not over finals. Propped up in the single bed of his darkened dorm room, he's masturbating rapidly with his right hand. With his left, he's using a one-handed keyboard to manipulate a Saran-wrapped laptop computer. As he idles through pictures of buxom naked women lolling around in various erotic poses, his heart pounds ever more loudly in his chest.

As he becomes increasingly excited, Roy adjusts the "arousal meter" on the computer screen upward. As he reaches the bright red "high" zone, a question pops up on the screen:

Could you enjoy sex with someone you hated?

Roy moves his left hand to a scale that ranges from "no" to "yes" and taps his answer. The next question appears: "Would you slip a woman a drug to increase the chance that she would have sex with you?"

Again, Roy selects his answer, and a new question pops up. "Would you always use a condom?"

BERKELEY ITSELF IS a dichotomous place. It was a site of antiestablishment riots in the 1960s, and people in the Bay Area snarkily refer to the famously left-of-center city as the "People's Republic of Berkeley." But the large campus itself draws a surprisingly conformist population of top-level students. In a survey of incoming freshmen in 2004, only 51.2 percent of the respondents thought of themselves as liberal. More than one-third (36 percent) deemed their views middle-of-the-road, and 12 percent claimed to be conservatives. To my surprise, when I arrived at Berkeley, I found that the students were in general not very wild, rebellious, or likely to take risks.

The ads we posted around Sproul Plaza read as follows: "Wanted: Male research participants, heterosexual, 18 years-plus, for a study on decision making and arousal." The ad noted that the experimental sessions would demand about an hour of the participants' time, that the participants would be paid $10 per session, and that the experiments could involve sexually arousing material. Those interested in applying could respond to Mike, the research assistant, by e-mail.

For this study, we decided to seek out only men. In terms of sex, their wiring is a lot simpler than that of women (as we concluded after much discussion among ourselves and our assistants, both male and female). A copy of *Playboy* and a darkened room were about all we'd need for a high degree of success.

Another concern was getting the project approved at MIT's Sloan School of Management (where I had my primary appointment). This was an ordeal in itself. Before allowing the research to begin, Dean Richard Schmalensee assigned a committee, consisting mostly of women, to examine the project. This committee had several concerns. What if a participant uncovered repressed memories of sexual abuse as a result of the research? Suppose a participant found that he or she was a sex addict? Their questions seemed unwarranted to me, since any college student with a computer and an Internet connection can get hold of the most graphic pornography imaginable.

Although the business school was stymied by this project, I was fortunate to have a position at MIT's Media Lab as well, and Walter Bender, who was the head of the lab, happily approved the project. I was on my way. But my experience with MIT's Sloan School made it clear that even half a century after Kinsey, and despite its substantial importance, sex is still largely a taboo subject for a study—at least at some institutions.

IN ANY CASE, our ads went out; and, college men being what they are, we soon had a long list of hearty fellows awaiting the chance to participate—including Roy.

Roy, in fact, was typical of most of the 25 participants in our study. Born and raised in San Francisco, he was accomplished, intelligent, and kind—the type of kid every prospective mother-in-law dreams of. Roy played Chopin études on the piano and liked to dance to techno music. He had earned straight A's throughout high school, where he was captain of the varsity volleyball team. He sympathized with libertarians and tended to vote Republican. Friendly and amiable, he had a steady girlfriend who he'd been dating for a year. He planned to go to medical school and had a weakness for spicy California-roll sushi and for the salads at Cafe Intermezzo.

Roy met with our student research assistant, Mike, at Strada coffee shop—Berkeley's patio-style percolator for many an intellectual thought, including the idea for the solution to Fermat's last theorem. Mike was slender and tall, with short hair, an artistic air, and an engaging smile.

Mike shook hands with Roy, and they sat down. "Thanks for answering our ad, Roy," Mike said, pulling out a few sheets of paper and placing them on the table. "First, let's go over the consent forms."

Mike intoned the ritual decree: The study was about decision making and sexual arousal. Participation was voluntary. Data would be confidential. Participants had the right to contact the committee in charge of protecting the rights of those participating in experiments, and so on.

Roy nodded and nodded. You couldn't find a more agreeable participant.

"You can stop the experiment at any time," Mike concluded. "Everything understood?"

"Yes," Roy said. He grabbed a pen and signed. Mike shook his hand.

"Great!" Mike took a cloth bag out of his knapsack. "Here's what's going to happen." He unwrapped an Apple iBook computer and opened it up. In addition to the standard keyboard, Roy saw a 12-key multicolored keypad.

"It's a specially equipped computer," Mike explained. "Please use only this keypad to respond." He touched the keys on the colored pad. "We'll give you a code to enter, and this code will let you start the experiment. During the session, you'll be asked a series of questions to which you can answer on a scale ranging between 'no' and 'yes.' If you think you would like the activity described in the question, answer 'yes,' and if you think you would not, answer 'no.' Remember that you're being asked to predict how you would behave and what kind of activities you would like when aroused."

Roy nodded.

"We'll ask you to sit in your bed, and set the computer up on a chair on the left side of your bed, in clear sight and reach of your bed," Mike went on. "Place the keypad next to you so that you can use it without any difficulty, and be sure you're alone."

Roy's eyes twinkled a little.

"When you finish with the session, e-mail me and we will meet again, and you'll get your ten bucks."

Mike didn't tell Roy about the questions themselves. The session started by asking Roy to imagine that he was sexually aroused, and to answer all the questions as he would if he were aroused. One set of questions asked about sexual preferences. Would he, for example, find women's shoes erotic? Could he imagine being attracted to a 50-year-old woman? Could it be fun to have sex with someone who was extremely fat? Could having sex with someone he hated be enjoyable? Would it be fun to get tied up or to tie someone else up? Could "just kissing" be frustrating?

A second set of questions asked about the likelihood of engaging in immoral behaviors such as date rape. Would Roy tell a woman that he loved her to increase the chance that she would have sex with him? Would he encourage a date to drink to increase the chance that she would have sex with him? Would he keep trying to have sex after a date had said "no"?

A third set of questions asked about Roy's likelihood of engaging in behaviors related to unsafe sex. Does a condom decrease sexual pleasure? Would he always use a condom if he didn't know the sexual history of a new sexual partner? Would he use a condom even if he was afraid that a woman might change her mind while he went to get it?*

A few days later, having answered the questions in his "cold," rational state, Roy met again with Mike.

"Those were some interesting questions," Roy noted.

"Yes, I know," Mike said coolly. "Kinsey had nothing on us. By the way, we have another set of experimental sessions. Would you be interested in participating again?"

Roy smiled a little, shrugged, and nodded.

Mike shoved a few pages toward him. "This time we're asking you to sign the same consent form, but the next task will be slightly different. The next session will be very much the same as the last one, but this time we want you to get yourself into an excited state by viewing a set of arousing pictures and masturbating. What we want you to do is arouse yourself to a high level, but not to ejaculate. In case you do, though, the computer will be protected."

Mike pulled out the Apple iBook. This time the keyboard and the screen were covered with a thin layer of Saran wrap.

Roy made a face. "I didn't know computers could get pregnant."

"Not a chance," Mike laughed. "This one had its tubes tied. But we like to keep them clean."

Mike explained that Roy would browse through a series

---

*For a complete lists of the questions we asked, see the appendix to this chapter.

of erotic pictures on the computer to help him get to the right level of arousal; then he would answer the same questions as before.

WITHIN THREE MONTHS, some fine Berkeley undergraduate students had undergone a variety of sessions in different orders. In the set of sessions conducted when they were in a cold, dispassionate state, they predicted what their sexual and moral decisions would be if they were aroused. In the set of sessions conducted when they were in a hot, aroused state, they also predicted their decisions—but this time, since they were actually in the grip of passion, they were presumably more aware of their preferences in that state. When the study was completed, the conclusions were consistent and clear—overwhelmingly clear, frighteningly clear.

In every case, our bright young participants answered the questions very differently when they were aroused from when they were in a "cold" state. Across the 19 questions about sexual preferences, when Roy and all the other participants were aroused they predicted that their desire to engage in a variety of somewhat odd sexual activities would be nearly twice as high as (72 percent higher than) they had predicted when they were cold. For example, the idea of enjoying contact with animals was more than twice as appealing when they were in a state of arousal as when they were in a cold state. In the five questions about their propensity to engage in immoral activities, when they were aroused they predicted their propensity to be more than twice as high as (136 percent higher than) they had predicted in the cold state. Similarly, in the set of questions about using condoms, and despite the warnings that had been hammered into them over the years about the importance of condoms, they were 25 percent more likely in the aroused state than in the cold state to predict that they would forego condoms. In all these cases they failed to predict the influence of arousal on their sexual preferences, morality, and approach to safe sex.

The results showed that when Roy and the other participants were in a cold, rational, superego-driven state, they respected women; they were not particularly attracted to the odd sexual activities we asked them about; they always took the moral high ground; and they expected that they would always use a condom. They thought that they understood themselves, their preferences, and what actions they were capable of. But as it turned out, they completely underestimated their reactions.

No matter how we looked at the numbers, it was clear that the magnitude of underprediction by the participants was substantial. Across the board, they revealed in their unaroused state that they themselves did not know what they were like once aroused. Prevention, protection, conservatism, and morality disappeared completely from the radar screen. They were simply unable to predict the degree to which passion would change them.*

IMAGINE WAKING UP one morning, looking in the mirror, and discovering that someone else—something alien but human—has taken over your body. You're uglier, shorter, hairier; your lips are thinner, your incisors are longer, your nails are filthy, your face is flatter. Two cold, reptilian eyes

---

*These results apply most directly to sexual arousal and its influence on who we are; but we can also assume that other emotional states (anger, hunger, excitement, jealousy, and so on) work in similar ways, making us strangers to ourselves.

gaze back at you. You long to smash something, rape someone. You are not you. You are a monster.

Beset by this nightmarish vision, Robert Louis Stevenson
screamed in his sleep in the early hours of an autumn morning in 1885. Immediately after his wife awoke him, he set to
work on what he called a "fine bogey tale"—*Dr. Jekyll and
Mr. Hyde*—in which he said, "Man is not truly one, but truly
two." The book was an overnight success, and no wonder.
The story captivated the imagination of Victorians, who were
fascinated with the dichotomy between repressive propriety—
represented by the mild-mannered scientist Dr. Jekyll—and
uncontrollable passion, embodied in the murderous Mr.
Hyde. Dr. Jekyll thought he understood how to control himself. But when Mr. Hyde took over, look out.

The story was frightening and imaginative, but it wasn't
new. Long before Sophocles's *Oedipus Rex* and Shakespeare's
Macbeth, the war between interior good and evil had been
the stuff of myth, religion, and literature. In Freudian terms,
each of us houses a dark self, an id, a brute that can unpredictably wrest control away from the superego. Thus a pleasant, friendly neighbor, seized by road rage, crashes his car
into a semi. A teenager grabs a gun and shoots his friends. A
priest rapes a boy. All these otherwise good people assume
that they understand themselves. But in the heat of passion,
suddenly, with the flip of some interior switch, everything
changes.

Our experiment at Berkeley revealed not just the old story
that we are all like Jekyll and Hyde, but also something
new—that every one of us, regardless of how "good" we are,
underpredicts the effect of passion on our behavior. In every
case, the participants in our experiment got it wrong. Even
the most brilliant and rational person, in the heat of passion,

seems to be absolutely and completely divorced from the person he thought he was. Moreover, it is not just that people make wrong predictions about themselves—their predictions are wrong by a large margin.

Most of the time, according to the results of the study, Roy is smart, decent, reasonable, kind, and trustworthy. His frontal lobes are fully functioning, and he is in control of his behavior. But when he's in a state of sexual arousal and the reptilian brain takes over, he becomes unrecognizable to himself.

Roy thinks he knows how he will behave in an aroused state, but his understanding is limited. He doesn't truly understand that as his sexual motivation becomes more intense, he may throw caution to the wind. He may risk sexually transmitted diseases and unwanted pregnancies in order to achieve sexual gratification. When he is gripped by passion, his emotions may blur the boundary between what is right and what is wrong. In fact, he doesn't have a clue to how consistently wild he really is, for when he is in one state and tries to predict his behavior in another state, he gets it wrong.

Moreover, the study suggested that our inability to understand ourselves in a different emotional state does not seem to improve with experience; we get it wrong even if we spend as much time in this state as our Berkeley students spend sexually aroused. Sexual arousal is familiar, personal, very human, and utterly commonplace. Even so, we all systematically underpredict the degree to which arousal completely negates our superego, and the way emotions can take control of our behavior.

WHAT HAPPENS, THEN, when our irrational self comes alive in an emotional place that we think is familiar but in fact is unfamiliar? If we fail to really understand ourselves, is it possible to somehow predict how we or others will behave when "out of our heads"—when we're really angry, hungry, frightened, or sexually aroused? Is it possible to do something about this?

The answers to these questions are profound, for they indicate that we must be wary of situations in which our Mr. Hyde may take over. When the boss criticizes us publicly, we might be tempted to respond with a vehement e-mail. But wouldn't we be better off putting our reply in the "draft" folder for a few days? When we are smitten by a sports car after a test-drive with the wind in our hair, shouldn't we take a break—and discuss our spouse's plan to buy a minivan— before signing a contract to buy the car?

Here are a few more examples of ways to protect ourselves from ourselves:

Safe Sex

Many parents and teenagers, while in a cold, rational, Dr. Jekyll state, tend to believe that the mere promise of abstinence—commonly known as "Just say no"—is sufficient protection against sexually transmitted diseases and unwanted pregnancies. Assuming that this levelheaded thought will prevail even when emotions reach the boiling point, the advocates of "just saying no" see no reason to carry a condom with them. But as our study shows, in the heat of passion, we are all in danger of switching from "Just say no" to "Yes!" in a heartbeat; and if no condom is available, we are likely to say yes, regardless of the dangers.

What does this suggest? First, widespread availability of condoms is essential. We should not decide in a cool state whether or not to bring condoms; they must be there just in case. Second, unless we understand how we might react in an emotional state, we will not be able to predict this transformation. For teenagers, this problem is most likely exacerbated, and thus sex education should focus less on the physiology and biology of the reproductive system, and more on strategies to deal with the emotions that accompany sexual arousal. Third, we must admit that carrying condoms and even vaguely understanding the emotional firestorm of sexual arousal may not be enough.

There are most likely many situations where teenagers simply won't be able to cope with their emotions. A better strategy, for those who want to guarantee that teenagers avoid sex, is to teach teenagers that they must walk away from the fire of passion before they are close enough to be drawn in. Accepting this advice might not be easy, but our results suggest that it is easier for them to fight temptation before it arises than after it has started to lure them in. In other words, avoiding temptation altogether is easier than overcoming it.

To be sure, this sounds a lot like the "Just say no" campaign, which urges teenagers to walk away from sex when tempted. But the difference is that "Just say no" assumes we can turn off passion at will, at any point, whereas our study shows this assumption to be false. If we put aside the debate on the pros and cons of teenage sex, what is clear is that if we want to help teenagers avoid sex, sexually transmitted diseases, and unwanted pregnancies, we have two strategies. Either we can teach them how to say no *before* any temptation takes hold, and before a situation becomes impossible to resist; or alternatively, we can get them prepared to deal with the consequences of saying yes in the heat of passion (by carrying a condom, for example). One thing is sure: if we don't teach our young people how to deal with sex when they are half out of their minds, we are not only fooling them; we're fooling ourselves as well. Whatever lessons we teach them, we need to help them understand that they will react differently when they are calm and cool from when their hormones are raging at fever pitch (and of course the same also applies to our own behavior).

Safe Driving

Similarly, we need to teach teenagers (and everyone else) not to drive when their emotions are at a boil. It's not just inexperience and hormones that make so many teenagers crash their own or their parents' cars. It's also the car full of laughing friends, with the CD player blaring at an adrenaline-pumping decibel level, and the driver's right hand searching for the french fries or his girlfriend's knee. Who's thinking about risk in that situation? Probably no one. A recent study found that a teenager driving alone was 40 percent more likely to get into an accident than an adult. But with one other teenager in the car, the percentage was twice that—and with a third teenager along for the ride, the percentage doubled again.[8]

To react to this, we need an intervention that does not rely on the premise that teenagers will remember how they wanted to behave while in a cold state (or how their parents wanted them to behave) and follow these guidelines even when they are in a hot state. Why not build into cars precautionary devices to foil teenagers' behavior? Such cars might be equipped with a modified OnStar system that the teenager and the parents configure in a cold state. If a car exceeds 65 miles per hour on the highway, or more than 40 miles per hour in a residential zone, for example, there will be consequences. If the car exceeds the speed limit or begins to make erratic turns, the radio might switch from 2Pac to Schumann's Second Symphony (this would slow most teenagers). Or the car might blast the air conditioning in winter, switch on the heat in summer, or automatically call Mom (a real downer if the driver's friends are present). With these substantial and immediate consequences in mind, then, the driver and his or her friends would realize that it's time for Mr. Hyde to move over and let Dr. Jekyll drive.

This is not at all far-fetched. Modern cars are already full of computers that control the fuel injection, the climate system, and the sound system. Cars equipped with OnStar are already linked to a wireless network. With today's technology, it would be a simple matter for a car to automatically call Mom.

Better Life Decisions

Not uncommonly, women who are pregnant for the first time tell their doctors, before the onset of labor, that they will refuse any kind of painkiller. The decision made in their cold state is admirable, but they make this decision when they can't imagine the pain that can come with childbirth (let alone the challenges of child rearing). After all is said and done, they may wish they'd gone for the epidural.

With this in mind, Sumi (my lovely wife) and I, readying ourselves for the birth of our first child, Amit, decided to test our mettle before making any decisions about using an epidural. To do this, Sumi plunged her hands into a bucket of ice for two minutes (we did this on the advice of our birth coach, who swore to us that the resulting pain would

103

be similar to the pain of childbirth), while I coached her breathing. If Sumi was unable to bear the pain of this experience, we figured, she'd probably want painkillers when she was going through the actual birth. After two minutes of holding her hands in the ice bucket, Sumi clearly understood the appeal of an epidural. During the birth itself, any ounce of love Sumi ever had for her husband was completely transferred to the anesthesiologist, who produced the epidural at the critical point. (With our second child, we made it to the hospital about two minutes before Neta was born, so Sumi did end up experiencing an analgesic-free birth after all.)

LOOKING FROM ONE emotional state to another is difficult. It's not always possible; and as Sumi learned it can be painful. But to make informed decisions we need to somehow experience and understand the emotional state we will be in at the other side of the experience. Learning how to bridge this gap is essential to making some of the important decisions of our lives.

It is unlikely that we would move to a different city without asking friends who live there how they like it, or even choose to see a film without reading some reviews. Isn't it strange that we invest so little in learning about both sides of ourselves? Why should we reserve this subject for psychology classes when failure to understand it can bring about repeated failures in so many aspects of our lives? We need to explore the two sides of ourselves; we need to understand the cold state and the hot state; we need to see how the gap between the hot and cold states benefits our lives, and where it leads us astray.

What did our experiments suggest? It may be that our models of human behavior need to be rethought. Perhaps there is no such thing as a fully integrated human being. We may, in fact, be an agglomeration of multiple selves. Although there is nothing much we can do to get our Dr. Jekyll to fully appreciate the strength of our Mr. Hyde, perhaps just being aware that we are prone to making the wrong decisions when gripped by intense emotion may help us, in some way, to apply our knowledge of our "Hyde" selves to our daily activities.

How can we try to force our "Hyde" self to behave better? This is what Chapter 6 is about.

A complete list of the questions we asked, with the mean response and percentage differences. Each question was presented on a visual-analog scale that stretched between "no" on the left (zero) to "possibly" in the middle (50) to "yes" on the right (100).

TABLE 1
RATE THE ATTRACTIVENESS OF DIFFERENT ACTIVITIES

Question	Nonaroused	Aroused	Difference, percent
Are women's shoes erotic?	42	65	55
Can you imagine being attracted to a 12-year-old girl?	23	46	100
Can you imagine having sex with a 40-year-old woman?	58	77	33
Can you imagine having sex with a 50-year-old woman?	28	55	96
Can you imagine having sex with a 60-year-old woman?	7	23	229
Can you imagine having sex with a man?	8	14	75
Could it be fun to have sex with someone who was extremely fat?	13	24	85
Could you enjoy having sex with someone you hated?	53	77	45
If you were attracted to a woman and she proposed a threesome with a man, would you do it?	19	34	79
Is a woman sexy when she's sweating?	56	72	29
Is the smell of cigarette smoke arousing?	13	22	69
Would it be fun to get tied up by your sexual partner?	63	81	29
Would it be fun to tie up your sexual partner?	47	75	60

TABLE 1 (continued)
RATE THE ATTRACTIVENESS OF DIFFERENT ACTIVITIES

Question	Nonaroused	Aroused	Difference, percent
Would it be fun to watch an attractive woman urinating?	25	32	28
Would you find it exciting to spank your sexual partner?	61	72	18
Would you find it exciting to get spanked by an attractive woman?	50	68	36
Would you find it exciting to have anal sex?	46	77	67
Can you imagine getting sexually excited by contact with an animal?	6	16	167
Is just kissing frustrating?	41	69	68

TABLE 2
RATE THE LIKELIHOOD OF ENGAGING IN IMMORAL BEHAVIORS LIKE DATE RAPE (A STRICT ORDER OF SEVERITY IS NOT IMPLIED)

Question	Nonaroused	Aroused	Difference, percent
Would you take a date to a fancy restaurant to increase your chance of having sex with her?	55	70	27
Would you tell a woman that you loved her to increase the chance that she would have sex with you?	30	51	70
Would you encourage your date to drink to increase the chance that she would have sex with you?	46	63	37
Would you keep trying to have sex after your date says "no"?	20	45	125
Would you slip a woman a drug to increase the chance that she would have sex with you?	5	26	420

TABLE 3
RATE YOUR TENDENCY TO USE, AND OUTCOMES
OF NOT USING, BIRTH CONTROL

Question	Nonaroused	Aroused	Difference, percent
Birth control is the woman's responsibility.	34	44	29
A condom decreases sexual pleasure.	66	78	18
A condom interferes with sexual spontaneity.	58	73	26
Would you always use a condom if you didn't know the sexual history of a new sexual partner?	88	69	22
Would you use a condom even if you were afraid that a woman might change her mind while you went to get it?	86	60	30

CHAPTER 6

The Problem of Procrastination and Self-Control

*Why We Can't Make Ourselves Do
What We Want to Do*

Onto the American scene, populated by big homes, big cars, and big-screen plasma televisions, comes another big phenomenon: the biggest decline in the personal savings rate since the Great Depression.

Go back 25 years, and double-digit savings rates were the norm. As recently as 1994 the savings rate was nearly five percent. But by 2006 the savings rate had fallen below zero—to negative one percent. Americans were not only not saving; they were spending more than they earned. Europeans do a lot better—they save an average of 20 percent. Japan's rate is 25 percent. China's is 50 percent. So what's up with America?

I suppose one answer is that Americans have succumbed to rampant consumerism. Go back to a home built before we had to have everything, for instance, and check out the size of the closets. Our house in Cambridge, Massachusetts, for example, was built in 1890. It has no closets whatsoever. Houses in the 1940s had closets barely big enough to stand in. The closet of the 1970s was a bit larger, perhaps deep enough for a fondue pot, a box of eight-track tapes, and a few disco dresses. But the closet of today is a different breed. "Walk-in closet" means that you can literally walk in for quite a distance. And no matter how deep these closets are, Americans have found ways to fill them right up to the closet door.

Another answer—the other half of the problem—is the recent explosion in consumer credit. The average American family now has six credit cards (in 2005 alone, Americans received 6 billion direct-mail solicitations for credit cards). Frighteningly, the average family debt on these cards is about $9,000; and seven in 10 households borrow on credit cards to cover such basic living expenses as food, utilities, and clothing.

So wouldn't it just be wiser if Americans learned to save, as in the old days, and as the rest of the world does, by diverting some cash to the cookie jar, and delaying some purchases until we can really afford them? Why can't we save part of our paychecks, as we know we should? Why can't we resist those new purchases? Why can't we exert some good old-fashioned self-control?

The road to hell, they say, is paved with good intentions. And most of us know what that's all about. We promise to save for retirement, but we spend the money on a vacation. We vow to diet, but we surrender to the allure of the dessert cart. We promise to have our cholesterol checked regularly, and then we cancel our appointment.

How much do we lose when our fleeting impulses deflect

us from our long-term goals? How much is our health affected by those missed appointments and our lack of exercise? How much is our wealth reduced when we forget our vow to save more and consume less? Why do we lose the fight against procrastination so frequently?

IN CHAPTER 5 we discussed how emotions grab hold of us and make us view the world from a different perspective. Procrastination (from the Latin *pro*, meaning *for*; and *cras*, meaning *tomorrow*) is rooted in the same kind of problem. When we promise to save our money, we are in a cool state. When we promise to exercise and watch our diet, again we're cool. But then the lava flow of hot emotion comes rushing in: just when we promise to save, we see a new car, a mountain bike, or a pair of shoes that we must have. Just when we plan to exercise regularly, we find a reason to sit all day in front of the television. And as for the diet? I'll take that slice of chocolate cake and begin the diet in earnest tomorrow. Giving up on our long-term goals for immediate gratification, my friends, is procrastination.

As a university professor, I'm all too familiar with procrastination. At the beginning of every semester my students make heroic promises to themselves—vowing to read their assignments on time, submit their papers on time, and in general, stay on top of things. And every semester I've watched as temptation takes them out on a date, over to the student union for a meeting, and off on a ski trip in the mountains—while their workload falls farther and farther behind. In the end, they wind up impressing me, not with their punctuality, but with their creativity—inventing stories, excuses, and family tragedies to explain their tardiness. (Why do family tragedies generally occur during the last two weeks of the semester?)

After I'd been teaching at MIT for a few years, my colleague Klaus Wertenbroch (a professor at INSEAD, a business school with campuses in France and Singapore) and I decided to work up a few studies that might get to the root of the problem, and just maybe offer a fix for this common human weakness. Our guinea pigs this time would be the delightful students in my class on consumer behavior.

As they settled into their chairs that first morning, full of anticipation (and, no doubt, with resolutions to stay on top of their class assignments), the students listened to me review the syllabus for the course. There would be three main papers over the 12-week semester, I explained. Together, these papers would constitute much of their final grade.

"And what are the deadlines?" asked one of them, waving his hand from the back. I smiled. "You can hand in the papers at any time before the end of the semester," I replied. "It's entirely up to you." The students looked back blankly.

"Here's the deal," I explained. "By the end of the week, you must commit to a deadline date for each paper. Once you set your deadlines, they can't be changed." Late papers, I added, would be penalized at the rate of one percent off the grade for each day late. The students could always turn in their papers before their deadlines without penalty, of course, but since I wouldn't be reading any of them until the end of the semester, there would be no particular advantage in terms of grades for doing so.

In other words, the ball was in their court. Would they have the self-control to play the game?

"But Professor Ariely," asked Gaurav, a clever master's student with a charming Indian accent, "given these instructions and incentives, wouldn't it make sense for us to select the last date possible?"

"You can do that," I replied. "If you find that it makes sense, by all means do it."

Under these conditions, what would you have done?

I promise to submit paper 1 on week ————
I promise to submit paper 2 on week ————
I promise to submit paper 3 on week ————

What deadlines did the students pick for themselves? A perfectly rational student would follow Gaurav's advice and set all the deadlines for the last day of class—after all, it was always possible to submit papers earlier without a penalty, so why take a chance and select an earlier deadline than needed? Delaying the deadlines to the end was clearly the best decision if students were perfectly rational. But what if the students are not rational? What if they succumb to temptation and are prone to procrastination? What if they realize their weakness? If the students are not rational, and they know it, they could use the deadlines to force themselves to behave better. They could set early deadlines and by doing so force themselves to start working on the projects earlier in the semester.

What did my students do? They used the scheduling tool I provided them with and spaced the timing of their papers across the whole semester. This is fine and good, as it suggests that the students realize their problems with procrastination and that if given the right opportunities they try to control themselves—but the main question is whether the tool was indeed helpful in improving their grades. To find out about this, we had to conduct other variations of the same experiments in other classes and compare the quality of papers across the different conditions (classes).

Now that I had Gaurav and his classmates choosing their individual deadlines, I went to my other two classes—with markedly different deals. In the second class, I told the students that they would have no deadlines at all during the semester. They merely needed to submit their papers by the end of the last class. They could turn the papers in early, of course, but there was no grade benefit to doing so. I suppose they should have been happy: I had given them complete flexibility and freedom of choice. Not only that, but they also had the lowest risk of being penalized for missing an intermediate deadline.

The third class received what might be called a dictatorial treatment: I dictated three deadlines for the three papers, set at the fourth, eighth, and twelfth weeks. These were my marching orders, and they left no room for choice or flexibility.

Of these three classes, which do you think achieved the best final grades? Was it Gaurav and his classmates, who had some flexibility? Or the second class, which had a single deadline at the end, and thus complete flexibility? Or the third class, which had its deadlines dictated from above, and therefore had no flexibility? Which class do you predict did worst?

When the semester was over, Jose Silva, the teaching assistant for the classes (himself an expert on procrastination and currently a professor at the University of California at Berkeley), returned the papers to the students. We could at last compare the grades across the three different deadline conditions. We found that the students in the class with the three firm deadlines got the best grades; the class in which I set no deadlines at all (except for the final deadline) had the worst grades; and the class in which Gaurav and his classmates were allowed to choose their own three deadlines (but with penalties for failing to meet them) finished in the middle, in terms of their grades for the three papers and their final grade.

What do these results suggest? First, that students do procrastinate (big news); and second, that tightly restricting their freedom (equally spaced deadlines, imposed from above) is the best cure for procrastination. But the biggest revelation is that simply offering the students a tool by which they could precommit to deadlines helped them achieve better grades.

What this finding implies is that the students generally understood their problem with procrastination and took action to fight it when they were given the opportunity to do so, achieving relative success in improving their grades. But why were the grades in the self-imposed deadlines condition not as good as the grades in the dictatorial (externally imposed) deadlines condition? My feeling is this: not everyone understands their tendency to procrastinate, and even those who do recognize their tendency to procrastinate may not understand their problem completely. Yes, people may set deadlines for themselves, but not necessarily the deadlines that are best for getting the best performance.

When I looked at the deadlines set by the students in Gaurav's class, this was indeed the case. Although the vast majority of the students in this class spaced their deadlines substantially (and got grades that were as good as those earned by students in the dictatorial condition), some did not space their deadlines much, and a few did not space their deadlines at all. These students who did not space their deadlines sufficiently pulled the average grades of this class down. Without properly spaced deadlines—deadlines that would have forced the students to start working on their papers earlier in the semester—the final work was generally rushed and poorly written (even without the extra penalty of one percent off the grade for each day of delay).

Interestingly, these results suggest that although almost everyone has problems with procrastination, those who recognize and admit their weakness are in a better position to utilize available tools for precommitment and by doing so, help themselves overcome it.

So THAT WAS my experience with my students. What does it have to do with everyday life? A lot, I think. Resisting temptation and instilling self-control are general human goals, and repeatedly failing to achieve them is a source of much of our misery. When I look around, I see people trying their best to do the right thing, whether they are dieters vowing to avoid a tempting dessert tray or families vowing to spend less and save more. The struggle for control is all around us. We see it in books and magazines. Radio and television airwaves are choked with messages of self-improvement and help.

And yet, for all this electronic chatter and focus in print, we find ourselves again and again in the same predicament as my students—failing over and over to reach our long-term goals. Why? Because without precommitments, we keep on falling for temptation.

What's the alternative? From the experiments that I have described above, the most obvious conclusion is that when an authoritative "external voice" gives the orders, most of us will jump to attention. After all, the students for whom I set the deadlines—for whom I provided the "parental" voice—did best. Of course, barking orders, while very effective, may not always be feasible or desirable. What's a good compromise? It seems that the best course might be to give people an opportunity to commit up front to their preferred path of action. This approach might not be as effective as the dictatorial treatment, but it can help push us in the right direction (perhaps even more so if we train people to do it, and give them experience in setting their own deadlines).

What's the bottom line? We have problems with self-control, related to immediate and delayed gratification—no doubt there. But each of the problems we face has potential self-control mechanisms, as well. If we can't save from our paycheck, we can take advantage of our employer's automatic deduction option; if we don't have the will to exercise regularly alone, we can make an appointment to exercise in the company of our friends. These are the tools that we can commit to in advance, and they may help us be the kind of people we want to be.

WHAT OTHER PROCRASTINATION problems might precommitment mechanisms solve? Consider health care and consumer debt.

Health Care

Everyone knows that preventive medicine is generally more cost-effective—for both individuals and society—than our current remedial approach. Prevention means getting health exams on a regular basis, before problems develop. But having a colonoscopy or mammogram is an ordeal. Even a cholesterol check, which requires blood to be drawn, is unpleasant. So while our long-term health and longevity depend on undergoing such tests, in the short term we procrastinate and procrastinate and procrastinate.

But can you imagine if we all got the required health exams on time? Think how many serious health problems could be caught if they were diagnosed early. Think how much cost could be cut from health-care spending, and how much misery would be saved in the process.

So how do we fix this problem? Well, we could have a dictatorial solution, in which the state (in the Orwellian sense) would dictate our regular checkups. That approach worked well with my students, who were given a deadline and performed well. In society, no doubt, we would all be healthier if the health police arrived in a van and took procrastinators to the ministry of cholesterol control for blood tests.

This may seem extreme, but think of the other dictates that society imposes on us for our own good. We may receive tickets for jaywalking, and for having our seat belts unsecured. No one thought 20 years ago that smoking would be banned in most public buildings across America, as well as in restaurants and bars, but today it is—with a hefty fine incurred for lighting up. And now we have the movement against trans fats. Should people be deprived of heart-clogging french fries?

Sometimes we strongly support regulations that restrain our self-destructive behaviors, and at other times we have equally strong feelings about our personal freedom. Either way, it's always a trade-off.

But if mandatory health checkups won't be accepted by the public, what about a middle ground, like the self-imposed deadlines I gave to Gaurav and his classmates (the deadlines that offered personal choice, but also had penalties attached for the procrastinators)? This might be the perfect compromise between authoritarianism, on the one hand, and what we have too often in preventive health today—complete freedom to fail.

Suppose your doctor tells you that you need to get your cholesterol checked. That means fasting the night before the blood test, driving to the lab the next morning without breakfast, sitting in a crowded reception room for what seems like hours, and finally, having the nurse come and get you so that she can stick a needle into your arm. Facing those prospects, you immediately begin to procrastinate. But suppose the doctor charged you an up-front $100 deposit for the test, refundable only if you showed up promptly at the appointed time. Would you be more likely to show up for the test?

What if the doctor asked you if you would like to pay this $100 deposit for the test? Would you accept this self-imposed challenge? And if you did, would it make you more likely to show up for the procedure? Suppose the procedure was more complicated: a colonoscopy, for instance. Would you be willing to commit to a $200 deposit, refundable only if you arrived at the appointment on time? If so, you will have replicated the condition that I offered Gaurav's class, a condition that certainly motivated the students to be responsible for their own decisions.

HOW ELSE COULD we defeat procrastination in health care? Suppose we could repackage most of our medical and dental procedures so that they were predictable and easily done. Let me tell you a story that illustrates this idea.

Several years ago, Ford Motor Company struggled to find

the best way to get car owners back into the dealerships for routine automobile maintenance. The problem was that the standard Ford automobile had something like 18,000 parts that might need servicing, and unfortunately they didn't all need servicing at the same time (one Ford engineer determined that a particular axle bolt needed inspection every 3,602 miles). And this was just part of the problem: since Ford had more than 20 vehicle types, plus various model years, the servicing of them all was nearly impossible to ponder. All that consumers, as well as service advisers, could do was page through volumes of thick manuals in order to determine what services were needed.

But Ford began to notice something over at the Honda dealerships. Even though the 18,000 or so parts in Honda cars had the same ideal maintenance schedules as the Ford cars, Honda had lumped them all into three "engineering intervals" (for instance, every six months or 5,000 miles, every year or 10,000 miles, and every two years or 25,000 miles). This list was displayed on the wall of the reception room in the service department. All the hundreds of service activities were boiled down to simple, mileage-based service events that were common across all vehicles and model years. The board had every maintenance service activity bundled, sequenced, and priced. Anyone could see when service was due and how much it would cost.

But the bundle board was more than convenient information: It was a true procrastination-buster, as it instructed customers to get their service done at specific times and mileages. It guided them along. And it was so simple that any customer could understand it. Customers were no longer confused. They no longer procrastinated. Servicing their Hondas on time was easy.

Some people at Ford thought this was a great idea, but at first the Ford engineers fought it. They had to be convinced that, yes, drivers could go 9,000 miles without an oil change—but that 5,000 miles would align the oil change with everything else that needed to be done. They had to be convinced that a Mustang and a F-250 Super Duty truck, despite their technological differences, could be put on the same maintenance schedule. They had to be convinced that rebundling their 18,000 maintenance options into three easily scheduled service events—making maintenance as easy as ordering a Value Meal at McDonald's—was not bad engineering, but good customer service (not to mention good business). The winning argument, in fact, was that it is better to have consumers service their vehicles at somewhat compromised intervals than not to service them at all!

In the end, it happened: Ford joined Honda in bundling its services. Procrastination stopped. Ford's service bay, which had been 40 percent vacant, filled up. The dealers made money, and in just three years Ford matched Honda's success in the service bay.

So couldn't we make comprehensive physicals and tests as simple—and, with the addition of self-imposed financial penalties (or better, a "parental" voice), bring the quality of our health way up and at the same time make the overall costs significantly less? The lesson to learn from Ford's experience is that bundling our medical tests (and procedures) so that people remember to do them is far smarter than adhering to an erratic series of health commands that people are unwilling to follow. And so the big question: can we shape America's medical morass and make it as easy as ordering a Happy Meal? Thoreau wrote, "Simplify! Simplify!" And, indeed, simplification is one mark of real genius.

Savings

We could order people to stop spending, as an Orwellian
edict. This would be similar to the case of my third group of
students, for whom the deadline was dictated by me. But are
there cleverer ways to get people to monitor their own spending? A few years ago, for instance, I heard about the "ice
glass" method for reducing credit card spending. It's a home
remedy for impulsive spending. You put your credit card into
a glass of water and put the glass in the freezer. Then, when
you impulsively decide to make a purchase, you must first
wait for the ice to thaw before extracting the card. By then,
your compulsion to purchase has subsided. (You can't just
put the card in the microwave, of course, because then you'd
destroy the magnetic strip.)

But here's another approach that is arguably better, and
certainly more up-to-date. John Leland wrote a very interesting article in the *New York Times* in which he described
a growing trend of self-shame: "When a woman who calls
herself Tricia discovered last week that she owed $22,302
on her credit cards, she could not wait to spread the news.
Tricia, 29, does not talk to her family or friends about her
finances, and says she is ashamed of her personal debt. Yet
from the laundry room of her home in northern Michigan,
Tricia does something that would have been unthinkable—
and impossible—a generation ago: She goes online and
posts intimate details of her financial life, including her net
worth (now a negative $38,691), the balance and finance
charges on her credit cards, and the amount of debt she has
paid down ($15,312) since starting the blog about her debt
last year."

It is also clear that Tricia's blog is part of a larger trend.
Apparently, there are dozens of Web sites (maybe there are

thousands by now) devoted to the same kind of debt blog-ging (from "Poorer than You" poorerthanyou.com and "We're in Debt" wereindebt.com to "Make Love Not Debt" makelovenotdebt.com and Tricia's Web page: blogging-awaydebt.com). Leland noted, "Consumers are asking oth-ers to help themselves develop self-control because so many companies are not showing any restraint."[9]

Blogging about overspending is important and useful, but as we saw in the last chapter, on emotions, what we truly need is a method to curb our consumption at the moment of temptation, rather than a way to complain about it after the fact.

What could we do? Could we create something that repli-cated the conditions of Gaurav's class, with some freedom of choice but built-in boundaries as well? I began to imagine a credit card of a different kind—a *self-control* credit card that would let people restrict their own spending behavior. The users could decide in advance how much money they wanted to spend in each category, in every store, and in every time frame. For instance, users could limit their spending on cof-fee to $20 every week, and their spending on clothing to $600 every six months. Cardholders could fix their limit for gro-ceries at $200 a week and their entertainment spending at $60 a month, and not allow any spending on candy between two and five PM. What would happen if they surpassed the limit? The cardholders would select their penalties. For in-stance, they could make the card get rejected; or they could tax themselves and transfer the tax to Habitat for Humanity, a friend, or long-term savings. This system could also imple-ment the "ice glass" method as a cooling-off period for large items; and it could even automatically trigger an e-mail to your spouse, your mother, or a friend:

> Dear Sumi,
> This e-mail is to draw your attention to the fact that
> your husband, Dan Ariely, who is generally an upright
> citizen, has exceeded his spending limit on chocolate of
> $50 per month by $73.25.
> With best wishes,
> The self-control credit card team

Now this may sound like a pipe dream, but it isn't. Think about the potential of Smart Cards (thin, palm-size cards that carry impressive computational powers), which are beginning to fill the market. These cards offer the possibility of being customized to each individual's credit needs and helping people manage their credit wisely. Why couldn't a card, for instance, have a spending "governor" (like the governors that limit the top speed on engines) to limit monetary transactions in particular conditions? Why couldn't they have the financial equivalent of a time-release pill, so that consumers could program their cards to dispense their credit to help them behave as they hope they would?

A FEW YEARS ago I was so convinced that a "self-control" credit card was a good idea that I asked for a meeting with one of the major banks. To my delight, this venerable bank responded, and suggested that I come to its corporate headquarters in New York.

I arrived in New York a few weeks later, and after a brief delay at the reception desk, was led into a modern conference room. Peering through the plate glass from on high, I could look down on Manhattan's financial district and a stream of yellow cabs pushing through the rain. Within a few minutes the room had filled with half a dozen high-powered banking executives, including the head of the bank's credit card division.

I began by describing how procrastination causes everyone problems. In the realm of personal finance, I said, it causes us to neglect our savings—while the temptation of easy credit fills our closets with goods that we really don't need. It didn't take long before I saw that I was striking a very personal chord with each of them.

Then I began to describe how Americans have fallen into a terrible dependence on credit cards, how the debt is eating them alive, and how they are struggling to find their way out of this predicament. America's seniors are one of the hardest-hit groups. In fact, from 1992 to 2004 the rate of debt of Americans age 55 and over rose faster than that of any other group. Some of them were even using credit cards to fill the gaps in their Medicare. Others were at risk of losing their homes.

I began to feel like George Bailey begging for loan forgiveness in *It's a Wonderful Life*. The executives began to speak up. Most of them had stories of relatives, spouses, and friends (not themselves, of course) who had had problems with credit debt. We talked it over.

Now the ground was ready and I started describing the self-control credit card idea as a way to help consumers spend less and save more. At first I think the bankers were a bit stunned. I was suggesting that they help consumers control their spending. Did I realize that the bankers and credit card companies made $17 billion a year in interest from these cards? Hello? They should give that up?

Well, I wasn't that naive. I explained to the bankers that there was a great business proposition behind the idea of a self-control card. "Look," I said, "the credit card business is cutthroat. You send out six billion direct-mail pieces a year, and all the card offers are about the same." Reluctantly, they agreed. "But suppose one credit card company stepped out of the pack," I continued, "and identified itself as a good guy—as an advocate for the credit-crunched consumer? Suppose one company had the guts to offer a card that would actually help consumers control their credit, and better still, divert some of their money into long-term savings?" I glanced around the room. "My bet is that thousands of consumers would cut up their other credit cards—and sign up with you!"

A wave of excitement crossed the room. The bankers nodded their heads and chatted to one another. It was revolutionary! Soon thereafter we all departed. They shook my hand warmly and assured me that we would be talking again, soon.

Well, they never called me back. (It might have been that they were worried about losing the $17 billion in interest charges, or maybe it was just good old procrastination.) But the idea is still there—a self-control credit card—and maybe one day someone will take the next step.

CHAPTER 7

The High Price
of Ownership

Why We Overvalue What We Have

At Duke University, basketball is somewhere between a passionate hobby and a religious experience. The basketball stadium is small and old and has bad acoustics—the kind that turn the cheers of the crowd into thunder and pump everyone's adrenaline level right through the roof. The small size of the stadium creates intimacy but also means there are not enough seats to contain all the fans who want to attend the games. This, by the way, is how Duke likes it, and the university has expressed little interest in exchanging the small, intimate stadium for a larger one. To ration the tickets, an intricate selection process has been developed over the years, to separate the truly devoted fans from all the rest.

Even before the start of the spring semester, students who want to attend the games pitch tents in the open grassy area outside the stadium. Each tent holds up to 10 students. The campers who arrive first take the spots closest to the stadium's entrance, and the ones who come later line up farther back. The evolving community is called Krzyzewskiville, reflecting the respect the students have for Coach K—Mike Krzyzewski—as well as their aspirations for victory in the coming season.

So that the serious basketball fans are separated from those without "Duke blue" running through their veins, an air horn is sounded at random times. At the sound, a countdown begins, and within the next five minutes at least one person from each tent must check in with the basketball authorities. If a tent fails to register within these five minutes, the whole tent gets bumped to the end of the line. This procedure continues for most of the spring semester, and intensifies in the last 48 hours before a game.

At that point, 48 hours before a game, the checks become "personal checks." From then on, the tents are merely a social structure: when the air horn is sounded, every student has to check in personally with the basketball authorities. Missing an "occupancy check" in these final two days can mean being bumped to the end of the line. Although the air horn sounds occasionally before routine games, it can be heard at all hours of night and day before the really big contests (such as games against the University of North Carolina-Chapel Hill and during the national championships).

But that's not the oddest part of the ritual. The oddest part is that for the really important games, such as the national titles, the students at the front of the line still don't get a ticket. Rather, each of them gets a lottery number. Only later, as they crowd around a list of winners posted at the student center, do they find out if they have really, truly won a ticket to the coveted game.

As Ziv Carmon (a professor at INSEAD) and I listened to the air horn during the campout at Duke in the spring of 1994, we were intrigued by the real-life experiment going on before our eyes. All the students who were camping out wanted passionately to go to the basketball game. They had all camped out for a long time for the privilege. But when the lottery was over, some of them would become ticket owners, while others would not.

The question was this: would the students who had won tickets—who had ownership of tickets—value those tickets more than the students who had not won them even though they all "worked" equally hard to obtain them? On the basis of Jack Knetsch, Dick Thaler, and Daniel Kahneman's research on the "endowment effect," we predicted that when we own something—whether it's a car or a violin, a cat or a basketball ticket—we begin to value it more than other people do.

Think about this for a minute. Why does the seller of a house usually value that property more than the potential buyer? Why does the seller of an automobile envision a higher price than the buyer? In many transactions why does the owner believe that his possession is worth more money than the potential owner is willing to pay? There's an old saying, "One man's ceiling is another man's floor." Well, when you're the owner, you're at the ceiling; and when you're the buyer, you're at the floor.

To be sure, that is not always the case. I have a friend who contributed a full box of record albums to a garage sale, for

predictably irrational

instance, simply because he couldn't stand hauling them
around any longer. The first person who came along offered
him $25 for the whole box (without even looking at the ti-
tles), and my friend accepted it. The buyer probably sold
them for 10 times that price the following day. Indeed, if we
always overvalued what we had, there would be no such
thing as *Antiques Roadshow*. ("How much did you pay for
this powder horn? Five dollars? Well, let me tell you, you
have a national treasure here.")

But this caveat aside, we still believed that in general the
ownership of something increases its value in the owner's
eyes. Were we right? Did the students at Duke who had won
the tickets—who could now anticipate experiencing the
packed stands and the players racing across the court—value
them more than the students who had not won them? There
was only one good way to find out: get them to tell us how
much they valued the tickets.

In this case, Ziv and I would try to buy tickets from some of
the students who had won them—and sell them to those who
didn't. That's right; we were about to become ticket scalpers.

THAT NIGHT WE got a list of the students who had won the
lottery and those who hadn't, and we started telephoning.
Our first call was to William, a senior majoring in chemistry.
William was rather busy. After camping for the previous
week, he had a lot of homework and e-mail to catch up on.
He was not too happy, either, because after reaching the
front of the line, he was still not one of the lucky ones who
had won a ticket in the lottery.

"Hi, William," I said. "I understand you didn't get one of
the tickets for the final four."

130

"That's right."

"We may be able to sell you a ticket."

"Cool."

"How much would you be willing to pay for one?"

"How about a hundred dollars?" he replied.

"Too low," I laughed. "You'll have to go higher."

"A hundred fifty?" he offered.

"You have to do better," I insisted. "What's the highest price you'll pay?"

William thought for a moment. "A hundred seventy-five."

"That's it?"

"That's it. Not a penny more."

"OK, you're on the list. I'll let you know," I said. "By the way, how'd you come up with that hundred seventy-five?"

William said he figured that for $175 he could also watch the game at a sports bar, free, spend some money on beer and food, and still have a lot left over for a few CDs or even some shoes. The game would no doubt be exciting, he said, but at the same time $175 is a lot of money.

Our next call was to Joseph. After camping out for a week Joseph was also behind on his schoolwork. But he didn't care—he had won a ticket in the lottery and now, in a few days, he would be watching the Duke players fight for the national title.

"Hi, Joseph," I said. "We may have an opportunity for you—to sell your ticket. What's your minimum price?"

"I don't have one."

"Everyone has a price," I replied, giving the comment my best Al Pacino tone.

His first answer was $3,000.

"Come on," I said, "That's way too much. Be reasonable; you have to offer a lower price."

"All right," he said, "twenty-four hundred."

"Are you sure?" I asked.

"That's as low as I'll go."

"OK. If I can find a buyer at that price, I'll give you a call. By the way," I added, "how did you come up with that price?"

"Duke basketball is a huge part of my life here," he said passionately. He then went on to explain that the game would be a defining memory of his time at Duke, an experience that he would pass on to his children and grandchildren. "So how can you put a price on that?" he asked. "Can you put a price on memories?"

William and Joseph were just two of more than 100 students whom we called. In general, the students who did not own a ticket were willing to pay around $170 for one. The price they were willing to pay, as in William's case, was tempered by alternative uses for the money (such as spending it in a sports bar for drinks and food). Those who owned a ticket, on the other hand, demanded about $2,400 for it. Like Joseph, they justified their price in terms of the importance of the experience and the lifelong memories it would create.

What was really surprising, though, was that in all our phone calls, not a single person was willing to sell a ticket at a price that someone else was willing to pay. What did we have? We had a group of students all hungry for a basketball ticket before the lottery drawing; and then, bang—in an instant after the drawing, they were divided into two groups—ticket owners and non–ticket owners. It was an emotional chasm that was formed, between those who now imagined the glory of the game, and those who imagined what else they could buy with the price of the ticket. And it was an empirical chasm as well—the average selling price (about $2,400) was separated by a factor of about 14 from the average buyer's offer (about $175).

From a rational perspective, both the ticket holders and the non–ticket holders should have thought of the game in exactly the same way. After all, the anticipated atmosphere at the game and the enjoyment one could expect from the experience should not depend on winning a lottery. Then how could a random lottery drawing have changed the students' view of the game—and the value of the tickets—so dramatically?

OWNERSHIP PERVADES OUR lives and, in a strange way, shapes many of the things we do. Adam Smith wrote, "Every man [and woman] . . . lives by exchanging, or becomes in some measure a merchant, and the society itself grows to be what is properly a commercial society." That's an awesome thought. Much of our life story can be told by describing the ebb and flow of our particular possessions—what we get and what we give up. We buy clothes and food, automobiles and homes, for instance. And we sell things as well— homes and cars, and in the course of our careers, our time.

Since so much of our lives is dedicated to ownership, wouldn't it be nice to make the best decisions about this? Wouldn't it be nice, for instance, to know exactly how much we would enjoy a new home, a new car, a different sofa, and an Armani suit, so that we could make accurate decisions about owning them? Unfortunately, this is rarely the case. We are mostly fumbling around in the dark. Why? Because of three irrational quirks in our human nature.

The first quirk, as we saw in the case of the basketball tickets, is that we fall in love with what we already have.

Suppose you decide to sell your old VW bus. What do you start doing? Even before you've put a FOR SALE sign in the window, you begin to recall trips you took. You were much younger, of course; the kids hadn't sprouted into teenagers. A warm glow of remembrance washes over you and the car. This applies not only to VW buses, of course, but to everything else. And it can happen fast.

For instance, two of my friends adopted a child from China and told me this remarkable story. They went to China with 12 other couples. When they reached the orphanage, the director took each of the couples separately into a room and presented them with a daughter. When the couples reconvened the following morning, they all commented on the director's wisdom: Somehow she knew exactly which little girl to give to each couple. The matches were perfect. My friends felt the same way, but they also realized that the matches had been random. What made each match seem perfect was not the Chinese woman's talent, but nature's ability to make us instantly attached to what we have.

The second quirk is that we focus on what we may lose, rather than what we may gain. When we price our beloved VW, therefore, we think more about what we will lose (the use of the bus) than what we will gain (money to buy something else). Likewise, the ticket holder focuses on losing the basketball experience, rather than imagining the enjoyment of obtaining money or on what can be purchased with it. Our aversion to loss is a strong emotion, and as I will explain later in the book, one that sometimes causes us to make bad decisions. Do you wonder why we often refuse to sell some of our cherished clutter, and if somebody offers to buy it, we attach an exorbitant price tag to it? As soon as we begin thinking about giving up our valued possessions, we are already mourning the loss.

The third quirk is that we assume other people will see the transaction from the same perspective as we do. We somehow expect the buyer of our VW to share our feelings, emotions, and memories. Or we expect the buyer of our house to appreciate how the sunlight filters through the kitchen windows. Unfortunately, the buyer of the VW is more likely to notice the puff of smoke that is emitted as you shift from first into second; and the buyer of your house is more likely to notice the strip of black mold in the corner. It is just difficult for us to imagine that the person on the other side of the transaction, buyer or seller, is not seeing the world as we see it.

OWNERSHIP ALSO HAS what I'd call "peculiarities." For one, the more work you put into something, the more ownership you begin to feel for it. Think about the last time you assembled some furniture. Figuring out which piece goes where and which screw fits into which hole boosts the feeling of ownership.

In fact, I can say with a fair amount of certainty that pride of ownership is inversely proportional to the ease with which one assembles the furniture; wires the high-definition television to the surround-sound system; installs software; or gets the baby into the bath, dried, powdered, diapered, and tucked away in the crib. My friend and colleague Mike Norton (a professor at Harvard) and I have a term for this phenomenon: the "Ikea effect."

Another peculiarity is that we can begin to feel ownership even before we own something. Think about the last time you entered an online auction. Suppose you make your first bid on Monday morning, for a wristwatch, and at this point you are the highest bidder. That night you log on, and you're still the top dog. Ditto for the next night. You start thinking about that elegant watch. You imagine it on your wrist; you imagine the compliments you'll get. And then you go online again one hour before the end of the auction. Some dog has topped your bid! Someone else will take your watch! So you increase your bid beyond what you had originally planned.

Is the feeling of partial ownership causing the upward spiral we often see in online auctions? Is it the case that the longer an auction continues, the greater grip virtual ownership will have on the various bidders and the more money they will spend? A few years ago, James Heyman, Yesim Orhun (a professor at the University of Chicago), and I set up an experiment to explore how the duration of an auction gradually affects the auction's participants and encourages them to bid to the bitter end. As we suspected, the participants who were the highest bidders, for the longest periods of time, ended up with the strongest feelings of virtual ownership. Of course, they were in a vulnerable position: once they thought of themselves as owners, they were compelled to prevent losing their position by bidding higher and higher.

"Virtual ownership," of course, is one mainspring of the advertising industry. We see a happy couple driving down the California coastline in a BMW convertible, and we imagine ourselves there. We get a catalog of hiking clothing from Patagonia, see a polyester fleece pullover, and—poof—we start thinking of it as ours. The trap is set, and we willingly walk in. We become partial owners even before we own anything.

There's another way that we can get drawn into ownership. Often, companies will have "trial" promotions. If we have a basic cable television package, for instance, we are lured into a "digital gold package" by a special "trial" rate (only $59 a month instead of the usual $89). After all, we tell ourselves, we can always go back to basic cable or downgrade to the "silver package."

But once we try the gold package, of course, we claim ownership of it. Will we really have the strength to downgrade back to basic or even to "digital silver"? Doubtful. At the onset, we may think that we can easily return to the basic service, but once we are comfortable with the digital picture, we begin to incorporate our ownership of it into our view of the world and ourselves, and quickly rationalize away the additional price. More than that, our aversion to loss—the loss of that nice crisp "gold package" picture and the extra channels—is too much for us to bear. In other words, before we make the switch we may not be certain that the cost of the digital gold package is worth the full price; but once we have it, the emotions of ownership come welling up, to tell us that the loss of "digital gold" is more painful than spending a few more dollars a month. We may think we can turn back, but that is actually much harder than we expected.

Another example of the same hook is the "30-day money-back guarantee." If we are not sure whether or not we should get a new sofa, the guarantee of being able to change our mind later may push us over the hump so that we end up getting it. We fail to appreciate how our perspective will shift once we have it at home, and how we will start viewing the sofa—as ours—and consequently start viewing returning it as a loss. We might think we are taking it home only to try it out for a few days, but in fact we are becoming owners of it and are unaware of the emotions the sofa can ignite in us.

OWNERSHIP IS NOT limited to material things. It can also apply to points of view. Once we take ownership of an idea—whether it's about politics or sports—what do we do? We love it perhaps more than we should. We prize it more than it is worth. And most frequently, we have trouble letting go of it because we can't stand the idea of its loss. What are we left with then? An ideology—rigid and unyielding.

THERE IS NO known cure for the ills of ownership. As Adam Smith said, it is woven into our lives. But being aware of it might help. Everywhere around us we see the temptation to improve the quality of our lives by buying a larger home, a second car, a new dishwasher, a lawn mower, and so on. But, once we change our possessions we have a very hard time going back down. As I noted earlier, ownership simply changes our perspective. Suddenly, moving backward to our pre-ownership state is a loss, one that we cannot abide. And so, while moving up in life, we indulge ourselves with the fantasy that we can always ratchet ourselves back if need be; but in reality, we can't. Downgrading to a smaller home, for instance, is experienced as a loss, it is psychologically painful, and we are willing to make all kinds of sacrifices in order to avoid such losses—even if, in this case, the monthly mortgage sinks our ship.

My own approach is to try to view all transactions (particularly large ones) as if I were a nonowner, putting some distance between myself and the item of interest. In this attempt, I'm not certain if I have achieved the uninterest in material things that is espoused by the Hindu sannyasi, but at least I try to be as Zen as I can about it.

CHAPTER 8

Keeping Doors Open

Why Options Distract Us from
Our Main Objective

In 210 BC, a Chinese commander named Xiang Yu led his troops across the Yangtze River to attack the army of the Qin (Ch'in) dynasty. Pausing on the banks of the river for the night, his troops awakened in the morning to find, to their horror, that their ships were burning. They hurried to their feet to fight off their attackers, but soon discovered that it was Xiang Yu himself who had set their ships on fire, and that he had also ordered all the cooking pots crushed.

Xiang Yu explained to his troops that without the pots and the ships, they had no other choice but to fight their way to victory or perish. That did not earn Xiang Yu a place on the Chinese army's list of favorite commanders, but it did have a tremendous focusing effect on his troops: grabbing their lances and bows, they charged ferociously against the enemy and won nine consecutive battles, completely obliterating the main-force units of the Qin dynasty.

Xiang Yu's story is remarkable because it is completely

antithetical to normal human behavior. Normally, we cannot stand the idea of closing the doors on our alternatives. Had most of us been in Xiang Yu's armor, in other words, we would have sent out part of our army to tend to the ships, just in case we needed them for retreat; and we would have asked others to cook meals, just in case the army needed to stay put for a few weeks. Still others would have been instructed to pound rice out into paper scrolls, just in case we needed parchment on which to sign the terms of the surrender of the mighty Qin (which was highly unlikely in the first place).

In the context of today's world, we work just as feverishly to keep all our options open. We buy the expandable computer system, just in case we need all those high-tech bells and whistles. We buy the insurance policies that are offered with the plasma high-definition television, just in case the big screen goes blank. We keep our children in every activity we can imagine—just in case one sparks their interest in gymnastics, piano, French, organic gardening, or tae kwon do. And we buy a luxury SUV, not because we really expect to drive off the highway, but because just in case we do, we want to have some clearance beneath our axles.

We might not always be aware of it, but in every case we give something up for those options. We end up with a computer that has more functions than we need, or a stereo with an unnecessarily expensive warranty. And in the case of our kids, we give up their time and ours—and the chance that they could become really good at one activity—in trying to give them some experience in a large range of activities. In running back and forth among the things that might be important, we forget to spend enough time on what really *is* important. It's a fool's game, and one that we are remarkably adept at playing.

I saw this precise problem in one of my undergraduate students, an extremely talented young man named Joe. As an incoming junior, Joe had just completed his required courses, and now he had to choose a major. But which one? He had a passion for architecture—he spent his weekends studying the eclectically designed buildings around Boston. He could see himself as a designer of such proud structures one day. At the same time he liked computer science, particularly the freedom and flexibility that the field offered. He could see himself with a good-paying job at an exciting company like Google. His parents wanted him to become a computer scientist—and besides, who goes to MIT to be an architect anyway?* Still, his love of architecture was strong.

As Joe spoke, he wrung his hands in frustration. The classes he needed for majors in computer science and architecture were incompatible. For computer science, he needed Algorithms, Artificial Intelligence, Computer Systems Engineering, Circuits and Electronics, Signals and Systems, Computational Structures, and a laboratory in Software Engineering. For architecture, he needed different courses: Experiencing Architecture Studio, Foundations in the Visual Arts, Introduction to Building Technology, Introduction to Design Computing, Introduction to the History and Theory of Architecture, and a further set of architecture studios.

How could he shut the door on one career or the other? If he started taking classes in computer science, he would have a hard time switching over to architecture; and if he started in architecture, he would have an equally difficult time switching to computer science. On the other hand, if he signed up for classes in both disciplines, he would most

---

* The architecture department at MIT is in fact very good.

likely end up without a degree in either field at the end of his four years at MIT, and he would require another year (paid for by his parents) to complete his degree. (He eventually graduated with a degree in computer science, but he found the perfect blend in his first job—designing nuclear subs for the Navy.)

Dana, another student of mine, had a similar problem—but hers centered on two boyfriends. She could dedicate her energy and passion to a person she had met recently and, she hoped, build an enduring relationship with him. Or she could continue to put time and effort into a previous relationship that was dying. She clearly liked the new boyfriend better than the former one—yet she couldn't let the earlier relationship go. Meanwhile, her new boyfriend was getting restless. "Do you really want to risk losing the boy you love," I asked her, "for the remote possibility that you may discover—at some later date—that you love your former boyfriend more?" She shook her head "no," and broke into tears.*

What is it about options that is so difficult for us? Why do we feel compelled to keep as many doors open as possible, even at great expense? Why can't we simply commit ourselves?†

To try to answer these questions, Jiwoong Shin (a professor at Yale) and I devised a series of experiments that we hoped would capture the dilemma represented by Joe and Dana. In our case, the experiment would be based on a com-

---

*I'm often surprised by how much people confide in me. I think it's partly due to my scars, and to the obvious fact that I've been through substantial trauma. On the other hand, what I would like to believe is that people simply recognize my unique insight into the human psyche, and thus seek my advice. Either way, I learn a lot from the stories people share with me.
†Matrimony is a social device that would seem to force individuals to shut down their alternative options, but, as we know, it too doesn't always work.

puter game that we hoped would eliminate some of the complexities of life and would give us a straightforward answer about whether people have a tendency to keep doors open for too long. We called it the "door game." For a location, we chose a dark, dismal place—a cavern that even Xiang Yu's army would have been reluctant to enter.

MIT's EAST CAMPUS dormitory is a daunting place. It is home to the hackers, hardware enthusiasts, oddballs, and general misfits (and believe me—it takes a serious misfit to be a misfit at MIT). One hall allows loud music, wild parties, and even public nudity. Another is a magnet for engineering students, whose models of everything from bridges to roller coasters can be found everywhere. (If you ever visit this hall, press the "emergency pizza" button, and a short time later a pizza will be delivered to you.) A third hall is painted completely black. A fourth has bathrooms adorned with murals of various kinds: press the palm tree or the samba dancer, and music, piped in from the hall's music server (all downloaded legally, of course), comes on.

One afternoon a few years ago, Kim, one of my research assistants, roamed the hallways of East Campus with a laptop tucked under her arm. At each door she asked the students whether they'd like to make some money participating in a quick experiment. When the reply was in the affirmative, Kim entered the room and found (sometimes only with difficulty) an empty spot to place the laptop.

As the program booted up, three doors appeared on the computer screen: one red, the second blue, and the third green. Kim explained that the participants could enter any of the three rooms (red, blue, or green) simply by clicking on the corresponding door. Once they were in a room, each subsequent click would earn them a certain amount of money. If a particular room offered between one cent and 10 cents, for instance, they would make something in that range each time they clicked their mouse in that room. The screen tallied their earnings as they went along.

Getting the most money out of this experiment involved finding the room with the biggest payoff and clicking in it as many times as possible. But this wasn't trivial. Each time you moved from one room to another, you used up one click (you had a total of 100 clicks). On one hand, switching from one room to another might be a good strategy for finding the biggest payout. On the other hand, running madly from door to door (and room to room) meant that you were burning up clicks which could otherwise have made you money.

Albert, a violin player (and a resident of the Dark Lord Krotus worshippers' hall), was one of the first participants. He was a competitive type, and determined to make more money than anyone else playing the game. For his first move, he chose the red door and entered the cube-shaped room.

Once inside, he clicked the mouse. It registered 3.5 cents. He clicked again; 4.1 cents; a third click registered one cent. After he sampled a few more of the rewards in this room, his interest shifted to the green door. He clicked the mouse eagerly and went in.

Here he was rewarded with 3.7 cents for his first click; he clicked again and received 5.8 cents; he received 6.5 cents the third time. At the bottom of the screen his earnings began to grow. The green room seemed better than the red room—but what about the blue room? He clicked to go through that last unexplored door. Three clicks fell in the range of four cents. Forget it. He hurried back to the green door (the room paying about five cents a click) and spent the remainder of his 100 clicks there, increasing his payoff. At the end, Albert inquired about his score. Kim smiled as she told him it was one of the best so far.

ALBERT HAD CONFIRMED something that we suspected about human behavior: given a simple setup and a clear goal (in this case, to make money), all of us are quite adept at pursuing the source of our satisfaction. If you were to express this experiment in terms of dating, Albert had essentially sampled one date, tried another, and even had a fling with a third. But after he had tried the rest, he went back to the best—and that's where he stayed for the remainder of the game.

But to be frank, Albert had it pretty easy. Even while he was running around with other "dates," the previous ones waited patiently for him to return to their arms. But suppose that the other dates, after a period of neglect, began to turn their backs on him? Suppose that his options began to close down? Would Albert let them go? Or would he try to hang on to all his options for as long as possible? In fact, would he sacrifice some of his guaranteed payoffs for the privilege of keeping these other options alive?

To find out, we changed the game. This time, any door left unvisited for 12 clicks would disappear forever.

SAM, A RESIDENT of the hackers' hall, was our first participant in the "disappearing" condition. He chose the blue door to begin with; and after entering it, he clicked three times. His earnings began building at the bottom of the screen, but this wasn't the only activity that caught his eye. With each additional click, the other doors diminished by one-twelfth, signifying that if not attended to, they would vanish. Eight more clicks and they would disappear forever.

Sam wasn't about to let that happen. Swinging his cursor around, he clicked on the red door, brought it up to its full size, and clicked three times inside the red room. But now he noticed the green door—it was four clicks from disappearing. Once again, he moved his cursor, this time restoring the green door to its full size.

The green door appeared to be delivering the highest payout. So should he stay there? (Remember that each room had a range of payouts. So Sam could not be completely convinced that the green door was actually the best. The blue might have been better, or perhaps the red, or maybe neither.) With a frenzied look in his eye, Sam swung his cursor across the screen. He clicked the red door and watched the blue door continue to shrink. After a few clicks in the red, he jumped over to the blue. But by now the green was beginning to get dangerously small—and so he was back there next.

Before long, Sam was racing from one option to another, his body leaning tensely into the game. In my mind I pictured a typically harried parent, rushing kids from one activity to the next.

Is this an efficient way to live our lives—especially when another door or two is added every week? I can't tell you the answer for certain in terms of your personal life, but in our experiments we saw clearly that running from pillar to post was not only stressful but uneconomical. In fact, in their frenzy to keep doors from shutting, our participants ended up making substantially less money (about 15 percent less) than the participants who didn't have to deal with closing doors. The truth is that they could have made more money by picking a room—any room—and merely staying there for the whole experiment! (Think about that in terms of your life or career.)

When Jiwoong and I tilted the experiments against keeping options open, the results were still the same. For instance, we made each click opening a door cost three cents, so that the cost was not just the loss of a click (an opportunity cost) but also a direct financial loss. There was no difference in response from our participants. They still had the same irrational excitement about keeping their options open.

Then we told the participants the exact monetary outcomes they could expect from each room. The results were still the same. They still could not stand to see a door close. Also, we allowed some participants to experience hundreds of practice trials before the actual experiment. Certainly, we thought, they would see the wisdom of *not* pursuing the closing doors. But we were wrong. Once they saw their options shrinking, our MIT students—supposedly among the best and brightest of young people—could not stay focused. Pecking like barnyard hens at every door, they sought to make more money, and ended up making far less.

In the end, we tried another sort of experiment, one that smacked of reincarnation. In this condition, a door would still disappear if it was not visited within 12 clicks. But it wasn't gone forever. Rather, a single click could bring it back to life. In other words, you could neglect a door without any loss. Would this keep our participants from clicking on it anyhow? No. To our surprise, they continued to waste their clicks on the "reincarnating" door, even though its disappearance had no real consequences and could always be easily reversed. They just couldn't tolerate the idea of the loss, and so they did whatever was necessary to prevent their doors from closing.

How CAN WE unshackle ourselves from this irrational impulse to chase worthless options? In 1941 the philosopher Erich Fromm wrote a book called *Escape from Freedom*. In a modern democracy, he said, people are beset not by a lack of opportunity, but by a dizzying abundance of it. In our modern society this is emphatically so. We are continually reminded that we can do anything and be anything we want to be. The problem is in living up to this dream. We must develop ourselves in every way possible; must taste every aspect of life; must make sure that of the 1,000 things to see before dying, we have not stopped at number 999. But then comes a problem—are we spreading ourselves too thin? The temptation Fromm was describing, I believe, is what we saw as we watched our participants racing from one door to another.

Running from door to door is a strange enough human activity. But even stranger is our compulsion to chase after doors of little worth—opportunities that are nearly dead, or that hold little interest for us. My student Dana, for instance, had already concluded that one of her suitors was most likely a lost cause. Then why did she jeopardize her relationship with the other man by continuing to nourish the wilting relationship with the less appealing romantic partner? Similarly, how many times have we bought something on sale not because we really needed it but because by the end of the sale all of those items would be gone, and we could never have it at that price again?

THE OTHER SIDE of this tragedy develops when we fail to realize that some things really are disappearing doors, and need our immediate attention. We may work more hours at our jobs, for instance, without realizing that the childhood of our sons and daughters is slipping away. Sometimes these doors close too slowly for us to see them vanishing. One of my friends told me, for instance, that the single best year of his marriage was when he was living in New York, his wife was living in Boston, and they met only on weekends. Before they had this arrangement—when they lived to-gether in Boston—they would spend their weekends catch-ing up on work rather than enjoying each other. But once the arrangement changed, and they knew that they had only the weekends together, their shared time became lim-ited and had a clear end (the time of the return train). Since it was clear that the clock was ticking, they dedi-cated the weekends to enjoying each other rather than do-ing their work.

I'm not advocating giving up work and staying home for the sake of spending all your time with your children, or moving to a different city just to improve your weekends with your spouse (although it might provide some benefits). But wouldn't it be nice to have a built-in alarm, to warn us when the doors are closing on our most important options?

SO WHAT CAN we do? In our experiments, we proved that running helter-skelter to keep doors from closing is a fool's game. It will not only wear out our emotions but also wear out our wallets. What we need is to consciously start closing some of our doors. Small doors, of course, are rather easy to close. We can easily strike names off our holiday card lists or omit the tae kwon do from our daughter's string of activities.

But the bigger doors (or those that seem bigger) are harder to close. Doors that just might lead to a new career or to a better job might be hard to close. Doors that are tied to our dreams are also hard to close. So are relationships with certain people—even if they seem to be going nowhere.

We have an irrational compulsion to keep doors open. It's just the way we're wired. But that doesn't mean we shouldn't try to close them. Think about a fictional episode: Rhett Butler leaving Scarlett O'Hara in *Gone with the Wind*, in the scene when Scarlett clings to him and begs him, "Where shall I go? What shall I do?" Rhett, after enduring too much from Scarlett, and finally having his fill of it, says, "Frankly, my dear, I don't give a damn." It's not by chance that this line has been voted the most memorable in cinematographic history. It's the emphatic closing of a door that gives it widespread appeal. And it should be a reminder to all of us that we have doors—little and big ones—which we ought to shut.

We need to drop out of committees that are a waste of our time and stop sending holiday cards to people who have moved on to other lives and friends. We need to determine whether we really have time to watch basketball and play both golf and squash and keep our family together; perhaps we should put some of these sports behind us. We ought to shut them because they draw energy and commitment away from the doors that should be left open—and because they drive us crazy.

SUPPOSE YOU'VE CLOSED so many of your doors that you have just two left. I wish I could say that your choices are easier now, but often they are not. In fact, choosing between two things that are similarly attractive is one of the most difficult decisions we can make. This is a situation not just of keeping options open for too long, but of being indecisive to the point of paying for our indecision in the end. Let me use the following story to explain.

A hungry donkey approaches a barn one day looking for hay and discovers two haystacks of identical size at the two opposite sides of the barn. The donkey stands in the middle of the barn between the two haystacks, not knowing which to select. Hours go by, but he still can't make up his mind. Unable to decide, the donkey eventually dies of starvation.*

This story is hypothetical, of course, and casts unfair aspersions on the intelligence of donkeys. A better example might be the U.S. Congress. Congress frequently gridlocks itself, not necessarily with regard to the big picture of particular legislation—the restoration of the nation's aging highways, immigration, improving federal protection of endangered species, etc.—but with regard to the details. Often, to a reasonable person, the party lines on these issues are the equivalent of the two bales of hay. Despite this, or because of it, Congress is frequently left stuck in the middle. Wouldn't a quick decision have been better for everybody?

---

* The French logician and philosopher Jean Buridan's commentaries on Aristotle's theory of action were the impetus of this story, known as "Buridan's ass."

Here's another example. One of my friends spent three months selecting a digital camera from two nearly identical models. When he finally made his selection, I asked him how many photo opportunities had he missed, how much of his valuable time he had spent making the selection, and how much he would have paid to have digital pictures of his family and friends documenting the last three months. More than the price of the camera, he said. Has something like this ever happened to you?

What my friend (and also the donkey and Congress) failed to do when focusing on the similarities and minor differences between two things was to take into account the *consequences of not deciding*. The donkey failed to consider starving, Congress failed to consider the lives lost while it debated highway legislation, and my friend failed to consider all the great pictures he was missing, not to mention the time he was spending at Best Buy. More important, they all failed to take into consideration the relatively minor differences that would have come with either one of the decisions.

My friend would have been equally happy with either camera; the donkey could have eaten either bale of hay; and the members of Congress could have gone home crowing over their accomplishments, regardless of the slight difference in bills. In other words, they all should have considered the decision an easy one. They could have even flipped a coin (figuratively, in the case of the donkey) and gotten on with their lives. But we don't act that way, because we just can't close those doors.

ALTHOUGH CHOOSING BETWEEN two very similar options should be simple, in fact it is not. I fell victim to this very same problem a few years ago, when I was considering whether to stay at MIT or move to Stanford (I chose MIT in the end). Confronted with these two options, I spent several weeks comparing the two schools closely and found that they were about the same in their overall attractiveness to me. So what did I do? At this stage of my problem, I decided I needed some more information and research on the ground. So I carefully examined both schools. I met people at each place and asked them how they liked it. I checked out neighborhoods and possible schools for our kids. Sumi and I pondered how the two options would fit in with the kind of life we wanted for ourselves. Before long, I was getting so engrossed in this decision that my academic research and productivity began to suffer. Ironically, as I searched for the best place to do my work, my research was being neglected.

Since you have probably invested some money to purchase my wisdom in this book (not to mention time, and the other activities you have given up in the process), I should probably not readily admit that I wound up like the donkey, trying to discriminate between two very similar bales of hay. But I did.

In the end, and with all my foreknowledge of the difficulty in this decision-making process, I was just as predictably irrational as everyone else.

CHAPTER 9

The Effect of Expectations

Why the Mind Gets What It Expects

Suppose you're a fan of the Philadelphia Eagles and you're watching a football game with a friend who, sadly, grew up in New York City and is a rabid fan of the Giants. You don't really understand why you ever became friends, but after spending a semester in the same dorm room you start liking him, even though you think he's football-challenged.

The Eagles have possession and are down by five points with no time-outs left. It's the fourth quarter, and six seconds are left on the clock. The ball is on the 12-yard line. Four wide receivers line up for the final play. The center hikes the ball to the quarterback who drops back in the pocket. As the receivers sprint toward the end zone, the quarterback throws a high pass just as the time runs out. An Eagles wide receiver near the corner of the end zone dives for the ball and makes a spectacular catch.

The referee signals a touchdown and all the Eagles players run onto the field in celebration. But wait. Did the receiver get both of his feet in? It looks close on the Jumbotron; so the booth calls down for a review. You turn to your friend: "Look at that! What a great catch! He was totally in. Why are they even reviewing it?" Your friend scowls. "That was completely out! I can't believe the ref didn't see it! You must be crazy to think that was in!"

What just happened? Was your friend the Giants fan just experiencing wishful thinking? Was he deceiving himself? Worse, was he lying? Or had his loyalty to his team—and his anticipation of its win—completely, truly, and deeply clouded his judgment?

I was thinking about that one evening, as I strolled through Cambridge and over to MIT's Walker Memorial Building. How could two friends—two honest guys—see one soaring pass in two different ways? In fact, how could any two parties look at precisely the same event and interpret it as supporting their opposing points of view? How could Democrats and Republicans look at a single schoolchild who is unable to read, and take such bitterly different positions on the same issue? How could a couple embroiled in a fight see the causes of their argument so differently?

A friend of mine who had spent time in Belfast, Ireland, as a foreign correspondent, once described a meeting he had arranged with members of the IRA. During the interview, news came that the governor of the Maze prison, a winding row of cell blocks that held many IRA operatives, had been assassinated. The IRA members standing around my friend, quite understandably, received the news with satisfaction—as a victory for their cause. The British, of course, didn't see it in those terms at all. The headlines in London the next day boiled with anger and calls for retribution. In fact, the British saw the event as proof that discussions with the IRA would lead nowhere and that the IRA should be crushed. I am an Israeli, and no stranger to such cycles of violence. Violence is not rare. It happens so frequently that we rarely stop to ask ourselves why. Why does it happen? Is it an outcome of history, or race, or politics—or is there something fundamentally irrational in us that encourages conflict, that causes us to look at the same event and, depending on our point of view, see it in totally different terms?

Leonard Lee (a professor at Columbia), Shane Frederick (a professor at MIT), and I didn't have any answers to these profound questions. But in a search for the root of this human condition, we decided to set up a series of simple experiments to explore how previously held impressions can cloud our point of view. We came up with a simple test—one in which we would not use religion, politics, or even sports as the indicator. We would use glasses of beer.

YOU REACH THE entrance to Walker by climbing a set of broad steps between towering Greek columns. Once inside (and after turning right), you enter two rooms with carpeting that predates the advent of electric light, furniture to match, and a smell that has the unmistaken promise of alcohol, packs of peanuts, and good company. Welcome to the Muddy Charles—one of MIT's two pubs, and the location for a set of studies that Leonard, Shane, and I would be conducting over the following weeks. The purpose of our experiments would be to determine whether people's expectations influence their views of subsequent events—more specifically, whether bar patrons' expectations for a certain kind of beer would shape their perception of its taste.

Let me explain this further. One of the beers that would be served to the patrons of the Muddy Charles would be Budweiser. The second would be what we fondly called MIT Brew. What's MIT Brew? Basically Budweiser, plus a "secret ingredient"—two drops of balsamic vinegar for each ounce of beer. (Some of the MIT students objected to our calling Budweiser "beer," so in subsequent studies, we used Sam Adams—a substance more readily acknowledged by Bostonians as "beer.")

At about seven that evening, Jeffrey, a second-year PhD student in computer science, was lucky enough to drop by the Muddy Charles. "Can I offer you two small, free samples of beer?" asked Leonard, approaching him. Without much hesitation, Jeffrey agreed, and Leonard led him over to a table that held two pitchers of the foamy stuff, one labeled A and the other B. Jeffrey sampled a mouthful of one of them, swishing it around thoughtfully, and then sampled the other. "Which one would you like a large glass of?" asked Leonard. Jeffrey thought it over. With a free glass in the offing, he wanted to be sure he would be spending his near future with the right malty friend.

Jeffrey chose beer B as the clear winner, and joined his friends (who were in deep conversation over the cannon that a group of MIT students had recently "borrowed" from the Caltech campus). Unbeknownst to Jeffrey, the two beers he had previewed were Budweiser and the MIT Brew—and the one he selected was the vinegar-laced MIT Brew.

A few minutes later, Mina, a visiting student from Estonia, dropped in. "Like a free beer?" asked Leonard. Her reply was a smile and a nod of the head. This time, Leonard offered more information. Beer A, he explained, was a standard commercial beer, whereas beer B had been doctored with a few drops of balsamic vinegar. Mina tasted the two beers. After finishing the samples (and wrinkling her nose at the vinegar-laced brew B) she gave the nod to beer A. Leonard poured her a large glass of the commercial brew and Mina happily joined her friends at the pub.

Mina and Jeffrey were only two of hundreds of students who participated in this experiment. But their reaction was typical: without foreknowledge about the vinegar, most of them chose the vinegary MIT Brew. But when they knew in advance that the MIT Brew had been laced with balsamic vinegar, their reaction was completely different. At the first taste of the adulterated suds, they wrinkled their noses and requested the standard beer instead. The moral, as you might expect, is that if you tell people up front that something might be distasteful, the odds are good that they will end up agreeing with you—not because their experience tells them so but because of their expectations.

If, at this point in the book, you are considering the establishment of a new brewing company, especially one that specializes in adding some balsamic vinegar to beer, consider the following points: (1) If people read the label, or knew about the ingredient, they would most likely hate your beer. (2) Balsamic vinegar is actually pretty expensive—so even if it makes beer taste better, it may not be worth the investment. Just brew a better beer instead.

BEER WAS JUST the start of our experiments. The MBA students at MIT's Sloan School also drink a lot of coffee. So one week, Elie Ofek (a professor at the Harvard Business School), Marco Bertini (a professor at the London Business School), and I opened an impromptu coffee shop, at which we offered students a free cup of coffee if they would answer a few questions about our brew. A line quickly formed. We handed our participants their cups of coffee and then pointed them to a table set with coffee additives—milk, cream, half-and-half, white sugar, and brown sugar. We also set out some unusual condiments—cloves, nutmeg, orange peel, anise, sweet paprika, and cardamom—for our coffee drinkers to add to their cups as they pleased.

After adding what they wanted (and none of our odd condiments were ever used) and tasting the coffee, the participants filled out a survey form. They indicated how much they liked the coffee, whether they would like it served in the cafeteria in the future, and the maximum price they would be willing to pay for this particular brew.

We kept handing out coffee for the next few days, but from time to time we changed the containers in which the odd condiments were displayed. Sometimes we placed them in beautiful glass-and-metal containers, set on a brushed metal tray with small silver spoons and nicely printed labels. At other times we placed the same odd condiments in white Styrofoam cups. The labels were handwritten in a red felt-tip pen. We went further and not only cut the Styrofoam cups shorter, but gave them jagged, hand-cut edges.

What were the results? No, the fancy containers didn't persuade any of the coffee drinkers to add the odd condiments (I guess we won't be seeing sweet paprika in coffee anytime soon). But the interesting thing was that when the odd condiments were offered in the fancy containers, the coffee drinkers were much more likely to tell us that they liked the coffee a lot, that they would be willing to pay well for it, and that they would recommend that we should start serving this new blend in the cafeteria. When the coffee ambience looked upscale, in other words, the coffee tasted upscale as well.

WHEN WE BELIEVE beforehand that something will be good, therefore, it generally will be good—and when we think it will be bad, it will bad. But how deep are these influences? Do they just change our beliefs, or do they also change the physiology of the experience itself? In other words, can previous knowledge actually modify the neural activity underlying the taste itself, so that when we expect something to taste good (or bad), it will actually taste that way?

To test this possibility, Leonard, Shane, and I conducted the beer experiments again, but with an important twist. We had already tested our MIT Brew in two ways—by telling our participants about the presence of vinegar in the beer *before* they tasted the brew, and by not telling them anything at all about it. But suppose we initially didn't tell them about the vinegar, then had them taste the beer, then revealed the presence of the vinegar, and then asked for their reactions. Would the placement of the knowledge—coming just after the experience—evoke a different response from what we received when the participants got the knowledge before the experience?

For a moment, let's switch from beer to another example. Suppose you heard that a particular sports car was fantastically exciting to drive, took one for a test drive, and then gave your impressions of the car. Would your impressions be different from those of people who didn't know anything about the sports car, took the test drive, then heard the car was hot, and then wrote down their impressions? In other words, does it matter if knowledge comes before or after the experience? And if so, which type of input is more important—knowledge before the experience, or an input of information after an experience has taken place?

The significance of this question is that if knowledge merely informs us of a state of affairs, then it shouldn't matter whether our participants received the information before or after tasting the beer: in other words, if we told them up front that there was vinegar in the beer, this should affect their review of the beer. And if we told them afterward, that should similarly affect their review. After all, they both got the same bad news about the vinegar-laced beer. This is what we should expect if knowledge merely *informs* us.

On the other hand, if telling our participants about the vinegar at the outset actually reshapes their sensory perceptions to align with this knowledge, then the participants who know about the vinegar up front should have a markedly different opinion of the beer from those who swigged a glass of it, and then were told. Think of it this way. If knowledge actually modifies the taste, then the participants who consumed the beer before they got the news about the vinegar, tasted the beer in the same way as those in the "blind" condition (who knew nothing about the vinegar). They learned about the vinegar only after their taste was established, at which point, if expectations change our experience, it was too late for the knowledge to affect the sensory perceptions.

So, did the students who were told about the vinegar after tasting the beer like it as little as the students who learned about the vinegar before tasting the beer? Or did they like it as much as the students who never learned about the vinegar? What do you think?

As it turned out, the students who found out about the vinegar after drinking the beer liked the beer much better than those who were told about the vinegar up front. In fact, those who were told afterward about the vinegar liked the beer just as much as those who weren't aware that there was any vinegar in the beer at all.

What does this suggest? Let me give you another example. Suppose Aunt Darcy is having a garage sale, trying to get rid of many things she collected during her long life. A car pulls up, some people get out, and before long they are gathered around one of the oil paintings propped up against the wall. Yes, you agree with them, it does look like a fine example of early American primitivism. But do you tell them that Aunt Darcy copied it from a photograph just a few years earlier?

My inclination, since I am an honest, upright person, would be to tell them. But should you tell them before or after they finish admiring the painting? According to our beer studies, you and Aunt Darcy would be better off keeping the information under wraps until after the examination. I'm not saying that this would entice the visitors to pay thousands of dollars for the painting (even though our beer drinkers preferred our vinegar-laced beer as much when they were told after drinking it as when they were not told at all), but it might get you a higher price for Aunt Darcy's work.

By the way, we also tried a more extreme version of this experiment. We told one of two groups in advance about the vinegar (the "before" condition) and told the second group about the vinegar after they had finished the sampling (the "after" condition). Once the tasting was done, rather than offer them a large glass of their choice, we instead gave them a large cup of unadulterated beer, some vinegar, a dropper, and the recipe for the MIT Brew (two drops of balsamic vinegar per ounce of beer). We wanted to see if people would freely add balsamic vinegar to their beer; if so, how much they would use; and how these outcomes would depend on whether the participants tasted the beer before or after knowing about the vinegar.

What happened? Telling the participants about the vinegar after rather than before they tasted the beer doubled the number of participants who decided to add vinegar to their beer. For the participants in the "after" condition, the beer with vinegar didn't taste too bad the first time around (they apparently reasoned), and so they didn't mind giving it another try.*

As YOU SEE, expectations can influence nearly every aspect of our life. Imagine that you need to hire a caterer for your daughter's wedding. Josephine's Catering boasts about its "delicious Asian-style ginger chicken" and its "flavorful Greek salad with kalamata olives and feta cheese." Another caterer, Culinary Sensations, offers a "succulent organic breast of chicken roasted to perfection and drizzled with a merlot demi-glace, resting in a bed of herbed Israeli couscous" and a "mélange of the freshest roma cherry tomatoes and crisp field greens, paired with a warm circle of chèvre in a fruity raspberry vinaigrette."

Although there is no way to know whether Culinary Sensations' food is any better than Josephine's, the sheer depth of the description may lead us to expect greater things from the simple tomato and goat cheese salad. This, accordingly, increases the chance that we (and our guests, if we give them the description of the dish) will rave over it.

This principle, so useful to caterers, is available to everyone. We can add small things that sound exotic and fashionable to our cooking (chipotle-mango sauces seem all the rage right now, or try buffalo instead of beef). These ingredients might not make the dish any better in a blind taste test; but

---

*We were also hoping to measure the amount of vinegar students added to the beer. But everyone who added vinegar added the exact amount specified in the recipe.

by changing our expectations, they can effectively influence the taste when we have this pre-knowledge.

These techniques are especially useful when you are inviting people for dinner—or persuading children to try new dishes. By the same token, it might help the taste of the meal if you omit the fact that a certain cake is made from a commercial mix or that you used generic rather than brand-name orange juice in a cocktail, or, especially for children, that Jell-O comes from cow hooves. I am not endorsing the morality of such actions, just pointing to the expected outcomes.

Finally, don't underestimate the power of presentation. There's a reason that learning to present food artfully on the plate is as important in culinary school as learning to grill and fry. Even when you buy take-out, try removing the Styrofoam packaging and placing the food on some nice dishes and garnishing it (especially if you have company); this can make all the difference.

One more piece of advice: If you want to enhance the experience of your guests, invest in a nice set of wineglasses.

Moreover, if you're really serious about your wine, you may want to go all out and purchase the glasses that are specific to burgundies, chardonnays, champagne, etc. Each type of glass is supposed to provide the appropriate environment, which should bring out the best in these wines (even though controlled studies find that the shape of the glass makes no difference at all in an objective blind taste test, that doesn't stop people from perceiving a significant difference when they are handed the "correct glass"). Moreover, if you forget that the shape of the glass really has no effect on the taste of the wine, you yourself may be able to better enjoy the wine you consume in the appropriately shaped fancy glasses.

Expectations, of course, are not limited to food. When you invite people to a movie, you can increase their enjoyment by mentioning that it got great reviews. This is also essential for building the reputation of a brand or product. That's what marketing is all about—providing information that will heighten someone's anticipated and real pleasure. But do expectations created by marketing really change our enjoyment?

I'm sure you remember the famous "Pepsi Challenge" ads on television (or at least you may have heard of them). The ads consisted of people chosen at random, tasting Coke and Pepsi and remarking about which they liked better. These ads, created by Pepsi, announced that people preferred Pepsi to Coke. At the same time, the ads for Coke proclaimed that people preferred Coke to Pepsi. How could that be? Were the two companies fudging their statistics?

The answer is in the different ways the two companies evaluated their products. Coke's market research was said to be based on consumers' preferences when they could see what they were drinking, including the famous red trademark, while Pepsi ran its challenge using blind tasting and standard plastic cups marked M and Q. Could it be that Pepsi tasted better in a blind taste test but that Coke tasted better in a non-blind (sighted) test?

To better understand the puzzle of Coke versus Pepsi, a terrific group of neuroscientists—Sam McClure, Jian Li, Damon Tomlin, Kim Cypert, Latané Montague, and Read Montague— conducted their own blind and non-blind taste test of Coke and Pepsi. The modern twist on this test was supplied by a functional magnetic resonance imaging (fMRI) machine. With this machine, the researchers could monitor the activity of the participants' brains while they consumed the drinks.

Tasting drinks while one is in an fMRI is not simple, by

the way, because a person whose brain is being scanned must lie perfectly still. To overcome this problem, Sam and his colleagues put a long plastic tube into the mouth of each participant, and from a distance injected the appropriate drink (Pepsi or Coke) through the tube into their mouths. As the participants received a drink, they were also presented with visual information indicating either that Coke was coming, that Pepsi was coming, or that an unknown drink was coming. This way the researchers could observe the brain activation of the participants while they consumed Coke and Pepsi, both when they knew which beverage they were drinking and when they did not.

What were the results? In line with the Coke and Pepsi "challenges," it turned out that the brain activation of the participants was different depending on whether the name of the drink was revealed or not. This is what happened: Whenever a person received a squirt of Coke or Pepsi, the center of the brain associated with strong feelings of emotional connection—called the ventromedial prefrontal cortex, VMPFC—was stimulated. But when the participants knew they were going to get a squirt of Coke, something additional happened. This time, the frontal area of the brain—the dorsolateral aspect of the prefrontal cortex, DLPFC, an area involved in higher human brain functions like working memory, associations, and higher-order cognitions and ideas—was also activated. It happened with Pepsi—but even more so with Coke (and, naturally, the response was stronger in people who had a stronger preference for Coke).

The reaction of the brain to the basic hedonic value of the drinks (essentially sugar) turned out to be similar for the two drinks. But the advantage of Coke over Pepsi was due to Cokes's brand—which activated the higher-order brain mechanisms.

These associations, then, and not the chemical properties of the drink, gave Coke an advantage in the marketplace.

It is also interesting to consider the ways in which the frontal part of the brain is connected to the pleasure center. There is a dopamine link by which the front part of the brain projects and activates the pleasure centers. This is probably why Coke was liked more when the brand was known—the associations were more powerful, allowing the part of the brain that represents these associations to enhance activity in the brain's pleasure center. This should be good news to any ad agency, of course, because it means that the bright red can, swirling script, and the myriad messages that have come down to consumers over the years (such as "Things go better with . . .") are as much responsible for our love of Coke as the brown bubbly stuff itself.

EXPECTATIONS ALSO SHAPE stereotypes. A stereotype, after all, is a way of categorizing information, in the hope of predicting experiences. The brain cannot start from scratch at every new situation. It must build on what it has seen before. For that reason, stereotypes are not intrinsically malevolent. They provide shortcuts in our never-ending attempt to make sense of complicated surroundings. This is why we have the expectation that an elderly person will need help using a computer or that a student at Harvard will be intelligent.* But because a stereotype provides us with specific expectations about members of a group, it can also unfavorably influence both our perceptions and our behavior.

---

*There is a nice T-shirt on sale at the MIT bookstore that reads "Harvard: Because not everyone can get into MIT."

Research on stereotypes shows not only that we react differently when we have a stereotype of a certain group of people, but also that stereotyped people themselves react differently when they are aware of the label that they are forced to wear (in psychological parlance, they are "primed" with this label). One stereotype of Asian-Americans, for instance, is that they are especially gifted in mathematics and science. A common stereotype of females is that they are weak in mathematics. This means that Asian-American women could be influenced by both notions.

In fact, they are. In a remarkable experiment, Margaret Shin, Todd Pittinsky, and Nalini Ambady asked Asian-American women to take an objective math exam. But first they divided the women into two groups. The women in one group were asked questions related to their gender. For example, they were asked about their opinions and preferences regarding coed dorms, thereby priming their thoughts for gender-related issues. The women in the second group were asked questions related to their race. These questions referred to the languages they knew, the languages they spoke at home, and their family's history in the United States, thereby priming the women's thoughts for race-related issues.

The performance of the two groups differed in a way that matched the stereotypes of both women and Asian-Americans. Those who had been reminded that they were women performed worse than those who had been reminded that they were Asian-American. These results show that even our own behavior can be influenced by our stereotypes, and that activation of stereotypes can depend on our current state of mind and how we view ourselves at the moment.

Perhaps even more astoundingly, stereotypes can also affect the behavior of people who are not even part of a stereotyped group. In one notable study, John Bargh, Mark Chen, and Lara Burrows had participants complete a scrambled-sentence task, rearranging the order of words to form sentences (we discussed this type of task in Chapter 4). For some of the participants, the task was based on words such as *aggressive, rude, annoying,* and *intrude.* For others, the task was based on words such as *honor, considerate, polite,* and *sensitive.* The goal of these two lists was to prime the participants to think about politeness or rudeness as a result of constructing sentences from these words (this is a very common technique in social psychology, and it works amazingly well).

After the participants completed the scrambled-sentence task, they went to another laboratory to participate in what was purportedly a second task. When they arrived at the second laboratory, they found the experimenter apparently in the midst of trying to explain the task to an uncomprehending participant who was just not getting it (this supposed participant was in fact not a real participant but a confederate working for the experimenter). How long do you think it took the real participants to interrupt the conversation and ask what they should do next?

The amount of waiting depended on what type of words had been involved in the scrambled-sentence task. Those who had worked with the set of polite words patiently waited for about 9.3 minutes before they interrupted, whereas those who had worked with the set of rude words waited only about 5.5 minutes before interrupting.

A second experiment tested the same general idea by priming the concept of the elderly, using words such as *Florida, bingo,* and *ancient.* After the participants in this experiment completed the scrambled-sentence task, they left the room, thinking that they had finished the experiment—but in fact the crux of the study was just beginning. What truly interested the researchers was how long it would take the participants to walk down the hallway as they left the building. Sure enough, the participants in the experimental group were affected by the "elderly" words: their walking speed was considerably slower than that of a control group who had not been primed. And remember, the primed participants were not themselves elderly people being reminded of their frailty—they were undergraduate students at NYU.

ALL THESE EXPERIMENTS teach us that expectations are more than the mere anticipation of a boost from a fizzy Coke. Expectations enable us to make sense of a conversation in a noisy room, despite the loss of a word here and there, and likewise, to be able to read text messages on our cell phones, despite the fact that some of the words are scrambled. And although expectations can make us look foolish from time to time, they are also very powerful and useful.

So what about our football fans and the game-winning pass? Although both friends were watching the same game, they were doing so through markedly different lenses. One saw the pass as in bounds. The other saw it as out. In sports, such arguments are not particularly damaging—in fact, they can be fun. The problem is that these same biased processes can influence how we experience other aspects of our world. These biased processes are in fact a major source of escalation in almost every conflict, whether Israeli-Palestinian, American-Iraqi, Serbian-Croatian, or Indian-Pakistani.

In all these conflicts, individuals from both sides can read similar history books and even have the same facts taught to them, yet it is very unusual to find individuals who would agree about who started the conflict, who is to blame, who should make the next concession, etc. In such matters, our investment in our beliefs is much stronger than any affiliation to sport teams, and so we hold on to these beliefs tenaciously. Thus the likelihood of agreement about "the facts" becomes smaller and smaller as personal investment in the problem grows. This is clearly disturbing. We like to think that sitting at the same table together will help us hammer out our differences and that concessions will soon follow. But history has shown us that this is an unlikely outcome; and now we know the reason for this catastrophic failure.

But there's reason for hope. In our experiments, tasting beer without knowing about the vinegar, or learning about the vinegar after the beer was tasted, allowed the true flavor to come out. The same approach should be used to settle arguments: The perspective of each side is presented without the affiliation—the facts are revealed, but not which party took which actions. This type of "blind" condition might help us better recognize the truth.

When stripping away our preconceptions and our previous knowledge is not possible, perhaps we can at least acknowledge that we are all biased. If we acknowledge that we are trapped within our perspective, which partially blinds us to the truth, we may be able to accept the idea that conflicts generally require a neutral third party—who has not been tainted with our expectations—to set down the rules and regulations. Of course, accepting the word of a third party is not easy and not always possible; but when it is possible, it can yield substantial benefits. And for that reason alone, we must continue to try.

CHAPTER 10

The Power of Price

Why a 50-Cent Aspirin Can Do What
a Penny Aspirin Can't

If you were living in 1950 and had chest pain, your cardiologist might well have suggested a procedure for angina pectoris called internal mammary artery ligation. In this operation, the patient is anesthetized, the chest is opened at the sternum, and the internal mammary artery is tied off. Voilà! Pressure to the pericardiophrenic arteries is raised, blood flow to the myocardium is improved, and everyone goes home happy.[10]

This was an apparently successful operation, and it had been a popular one for the previous 20 years. But one day in 1955, a cardiologist in Seattle, Leonard Cobb, and a few colleagues became suspicious. Was it really an effective procedure? Did it really work? Cobb decided to try to prove the efficacy of the procedure in a very bold way: he would perform the operation on half his patients, and fake the procedure on the other half. Then he would see which group felt better, and whose health actually improved. In other words, after 25 years of filleting patients like fish, heart surgeons would finally get a scientifically controlled surgical trial to see how effective the procedure really was.

To carry out this test, Dr. Cobb performed the traditional procedure on some of the patients, and placebo surgery on the others. The real surgery meant opening the patient up and tying up the internal mammary artery. In the placebo procedure, the surgeon merely cut into the patient's flesh with a scalpel, leaving two incisions. Nothing else was done.

The results were startling. Both the patients who did have their mammary arteries constricted and those who didn't reported immediate relief from their chest pain. In both groups, the relief lasted about three months—and then complaints about chest pain returned. Meanwhile, electrocardiograms showed no difference between those who had undergone the real operation and those who got the placebo operation. In other words, the traditional procedure seemed to provide some short-term relief—but so did the placebo. In the end, neither procedure provided significant long-term relief.

More recently a different medical procedure was submitted to a similar test, with surprisingly similar results. As early as 1993, J. B. Moseley, an orthopedic surgeon, had increasing doubts about the use of arthroscopic surgery for a particular arthritic affliction of the knee. Did the procedure really work? Recruiting 180 patients with osteoarthritis from the veterans' hospital in Houston, Texas, Dr. Moseley and his colleagues divided them into three groups.

One group got the standard treatment: anesthetic, three incisions, scopes inserted, cartilage removed, correction of soft-tissue problems, and 10 liters of saline washed through the knee. The second group got anesthesia, three incisions, scopes inserted, and 10 liters of saline, but no cartilage was removed. The third group—the placebo group—looked from the outside like the other two treatments (anesthesia, incisions, etc.); and the procedure took the same amount of time; but no instruments were inserted into the knee. In other words, this was simulated surgery.[11]

For two years following the surgeries, all three groups (which consisted of volunteers, as in any other placebo experiment) were tested for a lessening of their pain, and for the amount of time it took them to walk and climb stairs. How did they do? The groups that had the full surgery and the arthroscopic lavage were delighted, and said they would recommend the surgery to their families and friends. But strangely—and here was the bombshell—the placebo group also got relief from pain and improvements in walking—to the same extent, in fact, as those who had the actual operations. Reacting to this startling conclusion, Dr. Nelda Wray, one of the authors of the Moseley study, noted, "The fact that the effectiveness of arthroscopic lavage and debridement in patients with osteoarthritis of the knee is no greater than that of placebo surgery makes us question whether the $1 billion spent on these procedures might be put to better use."

If you assume that a firestorm must have followed this report, you're right. When the study appeared on July 11, 2002, as the lead article in the *New England Journal of Medicine*, some doctors screamed foul and questioned the method and results of the study. In response, Dr. Moseley argued that his study had been carefully designed and carried out. "Surgeons . . . who routinely perform arthroscopy are undoubtedly embarrassed at the prospect that the placebo effect—not surgical skill—is responsible for patient improvement after the surgeries they perform. As you might imagine, these surgeons are going to great lengths to try to discredit our study."

Regardless of the extent to which you believe the results of this study, it is clear that we should be more suspicious about arthroscopic surgery for this particular condition, and at the same time increase the burden of proof for medical procedures in general.

IN THE PREVIOUS chapter we saw that expectations change the way we perceive and appreciate experiences. Exploring the placebo effect in this chapter, we'll see not only that beliefs and expectations affect how we perceive and interpret sights, tastes, and other sensory phenomena, but also that our expectations can affect us by altering our subjective and even objective experiences—sometimes profoundly so.

Most important, I want to probe an aspect of placebos that is not yet fully understood. It is the role that *price* plays in this phenomenon. Does a pricey medicine make us feel better than a cheap medicine? Can it actually make us *physiologically* better than a cheaper brand? What about expensive procedures, and new-generation apparatuses, such as digital pacemakers and high-tech stents? Does their price influence their efficacy? And if so, does this mean that the bill for health care in America will continue to soar? Well, let's start at the beginning.

PLACEBO COMES FROM the Latin for "I shall please." The term was used in the fourteenth century to refer to sham mourners who were hired to wail and sob for the deceased at funerals. By 1785 it appeared in the *New Medical Dictionary*, attached to marginal practices of medicine.

One of the earliest recorded examples of the placebo effect in medical literature dates from 1794. An Italian physician named Gerbi made an odd discovery: when he rubbed the secretions of a certain type of worm on an aching tooth, the pain went away for a year. Gerbi went on to treat hundreds of patients with the worm secretions, keeping meticulous records of their reactions. Of his patients, 68 percent reported that their pain, too, went away for a year. We don't know the full story of Gerbi and his worm secretions, but we have a pretty good idea that the secretions really had nothing to do with curing toothaches. The point is that Gerbi believed they helped—and so did a majority of his patients.

Of course, Gerbi's worm secretion wasn't the only placebo in the market. Before recent times, almost all medicines were placebos. Eye of the toad, wing of the bat, dried fox lungs, mercury, mineral water, cocaine, an electric current: these were all touted as suitable cures for various ailments. When Lincoln lay dying across the street from Ford's Theater, it is said that his physician applied a bit of "mummy paint" to the wounds. Egyptian mummy, ground to a powder, was believed to be a remedy for epilepsy, abscesses, rashes, fractures, paralysis, migraine, ulcers, and many other things. As late as 1908, "genuine Egyptian mummy" could be ordered through the E. Merck catalog—and it's probably still in use somewhere today.[12]

Mummy powder wasn't the most macabre of medicines, though. One seventeenth-century recipe for a "cure all" medication advised: "Take the fresh corpse of a red-haired, uninjured, unblemished man, 24 years old and killed no more than one day before, preferably by hanging, breaking on the wheel or impaling. . . . Leave it one day and one night in the light of the sun and the moon, then cut into shreds or rough strips. Sprinkle on a little powder of myrrh and aloes, to prevent it from being too bitter."

We may think we're different now. But we're not. Placebos still work their magic on us. For years, surgeons cut remnants of scar tissue out of the abdomen, for instance, imagining that this procedure addressed chronic abdominal pain—until researchers faked the procedure in controlled studies and patients reported equal relief.[13] Encainide, flecainide, and mexiletine were widely prescribed off-label drugs for irregular heartbeat—and were later found to cause cardiac arrest.[14] When researchers tested the effect of the six leading antidepressants, they noted that 75 percent of the effect was duplicated in placebo controls.[15] The same was true of brain surgery for Parkinson's disease.[16] When physicians drilled holes in the skulls of several patients without performing the full procedure, to test its efficacy, the patients who received the sham surgery had the same outcome as those who received the full procedure. And of course the list goes on and on.

One could defend these modern procedures and compounds by noting that they were developed with the best intentions. This is true. But so were the applications of Egyptian mummy, to a great extent. And sometimes, the mummy powder worked just as well as (or at least no worse than) whatever else was used.

The truth is that placebos run on the power of suggestion. They are effective because people believe in them. You see your doctor and you feel better. You pop a pill and you feel better. And if your doctor is a highly acclaimed specialist, or your prescription is for a new wonder drug of some kind, you feel even better. But how does suggestion influence us?

IN GENERAL, TWO mechanisms shape the expectations that make placebos work. One is belief—our confidence or faith in the drug, the procedure, or the caregiver. Sometimes just the fact that a doctor or nurse is paying attention to us and reassuring us not only makes us feel better but also triggers our internal healing processes. Even a doctor's enthusiasm for a particular treatment or procedure may predispose us toward a positive outcome.

The second mechanism is conditioning. Like Pavlov's famous dogs (that learned to salivate at the ring of a bell), the body builds up expectancy after repeated experiences and releases various chemicals to prepare us for the future. Suppose you've ordered pizza night after night. When the deliveryman presses the doorbell, your digestive juices start flowing even before you can smell the pie. Or suppose that you are snuggled up on the couch with your loved one. As you're sitting there staring into a crackling fire, the prospect of sex releases endorphins, preparing you for what is to come next, and sending your sense of well-being into the stratosphere.

In the case of pain, expectation can unleash hormones and neurotransmitters, such as endorphins and opiates, that not only block agony but produce exuberant highs (endorphins trigger the same receptors as morphine). I vividly recall lying in the burn ward in terrible pain. As soon as I saw the nurse approaching, with a needle almost dripping with painkiller, what relief! My brain began secreting pain-dulling opioids, even before the needle broke my skin.

Thus familiarity may or may not breed contempt, but it definitely breeds expectations. Branding, packaging, and the reassurance of the caregiver can make us feel better. But what about price? Can the price of a drug also affect our response to it?

ON THE BASIS of price alone, it is easy to imagine that a $4,000 couch will be more comfortable than a $400 couch; that a pair of designer jeans will be better stitched and more comfortable than a pair from Wal-Mart; that a high-grade electric sander will work better than a low-grade sander; and that the roast duck at the Imperial Dynasty (for $19.95) is substantially better than the roast duck at Wong's Noodle Shop (for $10.95). But can such implied difference in quality influence the actual experience, and can such influence also apply to objective experiences such as our reactions to pharmaceuticals?

For instance, would a cheaper painkiller be less effective than a more expensive one? Would your winter cold feel worse if you took a discount cold medicine than if you took an expensive one? Would your asthma respond less well to a generic drug than to the latest brand-name on the market? In other words, are drugs like Chinese food, sofas, blue jeans, and tools? Can we assume that high price means higher quality, and do our expectations translate into the objective efficacy of the product?

This is a particularly important question. The fact is that you can get away with cheaper Chinese food and less expensive jeans. With some self-control, we can usually steer ourselves away from the most expensive brands. But will you really look for bargains when it comes to your health? Putting the common cold aside for the moment, are many of us going to pinch pennies when our lives are at risk? No—we want the best, for ourselves, our children, and our loved ones.

If we want the best for ourselves, does an expensive drug make us feel better than a cheaper drug? Does cost really make a difference in how we feel? In a series of experiments a few years ago, that's what Rebecca Waber (a graduate student at MIT), Baba Shiv (a professor at Stanford), Ziv Carmon, and I decided to find out.

IMAGINE THAT YOU'RE taking part in an experiment to test the efficacy of a new painkiller called Veladone-Rx. (The actual experiment involved about 100 adult Bostonians, but for now, we'll let you take their place.)

You arrive at the MIT Media Lab in the morning. Taya Leary, a young woman wearing a crisp business suit (this is in stark contrast to the usual attire of the students and faculty at MIT), greets you warmly, with a hint of a Russian accent. A photo ID identifies Taya as a representative of Vel Pharmaceuticals. She invites you to spend a moment reading a brochure about Veladone-Rx. Glancing around, you note that the room looks like a medical office: stale copies of *Time* and *Newsweek* are scattered around; brochures for Veladone-Rx are spread out on the table; and nearby is a cup of pens, with the drug's handsome logo. "Veladone is an exciting new medication in the opioid family," you read. "Clinical studies show that over 92 percent of patients receiving Veladone in double-blind controlled studies reported significant pain relief within only 10 minutes, and that pain relief lasted up to eight hours." And how much does it cost? According to the brochure, $2.50 for a single dose.

Once you finish reading the brochure, Taya calls in Rebecca Waber and leaves the room. Rebecca, wearing the white coat of a lab technician, with a stethoscope hanging from her neck, asks you a set of questions about your medical condition and your family's medical history. She listens to your heart and measures your blood pressure. Then she hooks you up to a complicated-looking machine. The electrodes running from the machine, greased with a green electrode gel, encircle your wrists. This is an electrical shock generator, she explains, and it is how we will test your perception and tolerance of pain.

With her hand on the switch, Rebecca sends a series of electrical shocks through the wires and into the electrodes. The initial shocks are merely annoying. Then they become painful, more painful, and finally so painful that your eyes fly open and your heart begins to race. She records your reactions. Now she starts delivering a new set of electrical shocks. This time she administers a set of charges that fluctuate randomly in intensity: some are very painful and some merely irritating. Following each one, you are asked to record, using the computer in front of you, the amount of pain you felt. You use the mouse to click on a line that ranges from "no pain at all" to "the worst pain imaginable" (this is called a "visual pain analog").

When this part of the torture ends, you look up. Rebecca is standing before you with a Veladone capsule in one hand and a cup of water in the other. "It will take about 15 minutes for the drug to reach its maximal effect," she says. You gulp it down, and then move to a chair in the corner, where you look at the old copies of *Time* and *Newsweek* until the pill takes effect.

Fifteen minutes later Rebecca, smearing the electrodes

with the same green electrode gel, cheerfully asks, "Ready for the next step?" You say nervously, "As ready as I can be." You're hooked up to the machine again, and the shocks begin. As before, you record the intensity of the pain after each shock. But this time it's different. It must be the Veladone-Rx! The pain doesn't feel nearly as bad. You leave with a pretty high opinion of Veladone. In fact, you hope to see it in the neighborhood drugstore before long.

Indeed, that's what most of our participants found. Almost all of them reported less pain when they experienced the electrical shocks under the influence of Veladone. Very interesting—considering that Veladone was just a capsule of vitamin C.

FROM THIS EXPERIMENT, we saw that our capsule did have a placebo effect. But suppose we priced the Veladone differently. Suppose we discounted the price of a capsule of Veladone-Rx from $2.50 to just 10 cents. Would our participants react differently?

In our next test, we changed the brochure, scratching out the original price ($2.50 per pill) and inserting a new discount price of 10 cents. Did this change our participants' reaction? Indeed. At $2.50 almost all our participants experienced pain relief from the pill. But when the price was dropped to 10 cents, only half of them did.

Moreover, it turns out that this relationship between price and placebo effect was not the same for all participants, and the effect was particularly pronounced for people who had more experience with recent pain. In other words, for people who had experienced more pain, and thus depended more on pain medications, the relationship was more pronounced:

they got even less benefit when the price was discounted. When it comes to medicines, then, we learned that you get what you pay for. Price can change the experience.

INCIDENTALLY, WE GOT corroborating results in another test, a study we conducted one miserably cold winter at the University of Iowa. In this case we asked a group of students to keep track of whether they used full-price or discount medicines for their seasonal colds, and if so, how well those remedies worked. At the end of the semester, 13 participants said they'd paid list price and 16 had bought discount drugs. Which group felt better? I think you can guess by now: the 13 who paid the list price reported significantly better medical outcomes than the 16 who bought the medication at a discount. And so, in over-the-counter cold medication, what you pay is often what you get.

FROM OUR EXPERIMENTS with our "pharmaceuticals" we saw how prices drive the placebo effect. But do prices affect everyday consumer products as well? We found the perfect subject in SoBe Adrenaline Rush, a beverage that promises to "elevate your game" and impart "superior functionality."

In our first experiment, we stationed ourselves at the entrance of the university's gym, offering SoBe. The first group of students paid the regular price for the drink. A second group also purchased the drink, but for them the price was marked down to about one-third of the regular price. After the students exercised, we asked them if they felt more or less fatigued relative to how they normally felt after their usual workouts. Both groups of students who drank the SoBe indicated that they were somewhat less fatigued than usual. That seemed plausible, especially considering the hefty shot of caffeine in each bottle of SoBe.

But it was the effect of the price, not the effect of the caffeine, that we were after. Would higher-priced SoBe reduce fatigue better than the discounted SoBe? As you can imagine from the experiment with Veladone, it did. The students who drank the higher-priced beverage reported less fatigue than those who had the discounted drink.

These results were interesting, but they were based on the participants' impressions of their own state—their subjective reports. How could we test SoBe more directly and objectively? We found a way: SoBe claims to provide "energy for your mind." So we decided to test that claim by using a series of anagrams.

It would work like this. Half of the students would buy their SoBe at full price, and the other half would buy it at a discount. (We actually charged their student accounts, so in fact their parents were the ones paying for it.) After consuming the drinks, the students would be asked to watch a movie for 10 minutes (to allow the effects of the beverage to sink in, we explained). Then we would give each of them a 15-word puzzle, with 30 minutes to solve as many of the problems as they could. (For example, when given the set TUPPIL, participants had to rearrange it to PULPIT—or they would have to rearrange FRIVEY, RENCOR, and SVALIE to get . . .).

We had already established a baseline, having given the word-puzzle test to a group of students who had not drunk SoBe. This group got on average nine of the 15 items right.

What happened when we gave the puzzles to the students who drank SoBe? The students who had bought it at the full price also got on average about nine answers right—this was no different from the outcome for those who had no drink at all. But more interesting were the answers from the discounted SoBe group: they averaged 6.5 questions right. What can we gather from this? Price does make a difference, and in this case the difference was a gap of about 28 percent in performance on the word puzzles.

So SoBe didn't make anyone smarter. Does this mean that the product itself is a dud (at least in terms of solving word puzzles)? To answer this question, we devised another test. The following message was printed on the cover of the quiz booklet: "Drinks such as SoBe have been shown to improve mental functioning," we noted, "resulting in improved performance on tasks such as solving puzzles." We also added some fictional information, stating that SoBe's Web site referred to more than 50 scientific studies supporting its claims.

What happened? The group that had the full-price drinks still performed better than those that had the discounted drinks. But the message on the quiz booklet also exerted some influence. Both the discount group and the full-price group, having absorbed the information and having been primed to expect success, did better than the groups whose quiz cover didn't have the message. And this time the SoBe did make people smarter. When we hyped the drink by stating that 50 scientific studies found SoBe to improve mental functioning, those who got the drink at the discount price improved their score (in answering additional questions) by 0.6, but those who got both the hype and the full price improved by 3.3 additional questions. In other words, the message on the bottle (and the quiz cover) as well as the price was arguably more powerful than the beverage inside.

ARE WE DOOMED, then, to get lower benefits every time we get a discount? If we rely on our irrational instincts, we will. If we see a discounted item, we will instinctively assume that its quality is less than that of a full-price item—and then in fact we will make it so. What's the remedy? If we stop and rationally consider the product versus the price, will we be able to break free of the unconscious urge to discount quality along with price?

We tried this in a series of experiments, and found that consumers who stop to reflect about the relationship between price and quality are far less likely to assume that a discounted drink is less effective (and, consequently, they don't perform as poorly on word puzzles as they would if they did assume it). These results not only suggest a way to overcome the relationship between price and the placebo effect but also suggest that the effect of discounts is largely an unconscious reaction to lower prices.

SO WE'VE SEEN how pricing drives the efficacy of placebo, painkillers, and energy drinks. But here's another thought. If placebos can make us feel better, should we simply sit back and enjoy them? Or are placebos patently bad—shams that should be discarded, whether they make us feel good or not? Before you answer this question, let me raise the ante. Suppose you found a placebo substance or a placebo procedure that not only made you feel better but actually made you physically better. Would you still use it? What if you were a physician? Would you prescribe medications that were only placebos? Let me tell you a story that helps explain what I'm suggesting.

In AD 800, Pope Leo III crowned Charlemagne emperor of the Romans, thus establishing a direct link between church and state. From then on the Holy Roman emperors, followed by the kings of Europe, were imbued with the glow of divinity. Out of this came what was called the "royal touch"—the practice of healing people. Throughout the Middle Ages, as one historian after another chronicled, the great kings would regularly pass through the crowds, dispensing the royal touch. Charles II, who ruled England from 1660 to 1685, for instance, was said to have touched some 100,000 people during his reign; and the records even include the names of several American colonists, who returned to the Old World from the New World just to cross paths with King Charles and be healed.

Did the royal touch really work? If no one had ever gotten better after receiving the royal touch, the practice would obviously have withered away. But throughout history, the royal touch was said to have cured thousands of people. Scrofula, a disfiguring and socially isolating disease often mistaken for leprosy, was believed to be dispelled by the royal touch. Shakespeare wrote in *Macbeth*: "Strangely visited people, All sworn and ulcerous, pitiful to the eye . . . Put on with holy prayers and 'tis spoken, the healing benediction." The royal touch continued until the 1820s, by which time monarchs were no longer considered heaven-sent—and (we might imagine) "new, improved!" advances in Egyptian mummy ointments made the royal touch obsolete.

When people think about a placebo such as the royal touch, they usually dismiss it as "just psychology." But, there is nothing "just" about the power of a placebo, and in reality it represents the amazing way our mind controls our body. How the mind achieves these amazing outcomes is not always very clear.* Some of the effect, to be sure, has to do with reducing the level of stress, changing hormonal secretions, changing the immune system, etc. The more we understand the connection between brain and body, the more things that once seemed clear-cut become ambiguous. Nowhere is this as apparent as with the placebo.

In reality, physicians provide placebos all the time. For instance, a study done in 2003 found that more than one-third of patients who received antibiotics for a sore throat were later found to have viral infections, for which an antibiotic does absolutely no good (and possibly contributes to the rising number of drug-resistant bacterial infections that threaten us all[17]). But do you think doctors will stop handing us antibiotics when we have viral colds? Even when doctors know that a cold is viral rather than bacterial (and many colds are viral), they still know very well that the patient wants some sort of relief; most commonly, the patient expects to walk out with a prescription. Is it right for the physician to fill this psychic need?

The fact that physicians give placebos all the time does not mean that they want to do this, and I suspect that the practice tends to make them somewhat uncomfortable. They've been trained to see themselves as men and women of science, people who must look to the highest technologies of modern medicine for answers. They want to think of themselves as real healers, not practitioners of voodoo. So it can

---

*We do understand quite precisely how a placebo works in the domain of pain, and this is why we selected the painkiller as our object of investigation. But other placebo effects are not as well understood.

be extremely difficult for them to admit, even to themselves, that their job may include promoting health through the placebo effect. Now suppose that a doctor does allow, however grudgingly, that a treatment he knows to be a placebo helps some patients. Should he enthusiastically prescribe it? After all, the physician's enthusiasm for a treatment can play a real role in its efficacy.

Here's another question about our national commitment to health care. America already spends more of its GDP per person on health care than any other Western nation. How do we deal with the fact that expensive medicine (the 50-cent aspirin) may make people feel better than cheaper medicine (the penny aspirin). Do we indulge people's irrationality, thereby raising the costs of health care? Or do we insist that people get the cheapest generic drugs (and medical procedures) on the market, regardless of the increased efficacy of the more expensive drugs? How do we structure the cost and co-payment of treatments to get the most out of medications, and how can we provide discounted drugs to needy populations without giving them treatments that are less effective? These are central and complex issues for structuring our health care system. I don't have the answers to these questions, but they are important for all of us to understand.

Placebos pose dilemmas for marketers, too. Their profession requires them to create perceived value. Hyping a product beyond what can be objectively proved is—depending on the degree of hype—stretching the truth or outright lying. But we've seen that the perception of value, in medicine, soft drinks, drugstore cosmetics, or cars, can become real value. If people actually get more satisfaction out of a product that has been hyped, has the marketer done anything worse than sell the sizzle along with the steak? As we start thinking more about placebos and the blurry boundary between beliefs and reality, these questions become more difficult to answer.

As a SCIENTIST I value experiments that test our beliefs and the efficacy of different treatments. At the same time, it is also clear to me that experiments, particularly those involving medical placebos, raise many important ethical questions. Indeed, the experiment involving mammary ligation that I mentioned at the beginning of this chapter raised an ethical issue: there was an outcry against performing sham operations on patients.

The idea of sacrificing the well-being and perhaps even the life of some individuals in order to learn whether a particular procedure should be used on other people at some point in the future is indeed difficult to swallow. Visualizing a person getting a placebo treatment for cancer, for example, just so that years later other people will perhaps get better treatment seems a strange and difficult trade-off to make.

At the same time, the trade-offs we make by *not* carrying out enough placebo experiments are also hard to accept. And as we have seen, they can result in hundreds or thousands of people undergoing useless (but risky) operations. In the United States very few surgical procedures are tested scientifically. For that reason, we don't really know whether many operations really offer a cure, or whether, like many of their predecessors, they are effective merely because of their placebo effect. Thus, we may find ourselves frequently submitting to procedures and operations that if more carefully studied, would be put aside. Let me share with you my own story of a procedure that, in my case, was highly touted, but in reality was nothing more than a long, painful experience.

I had been in the hospital for two long months when my oc-
cupational therapist came to me with exciting news. There
was a technological garment for people like me called the
Jobst suit. It was skinlike, and it would add pressure to what
little skin I had left, so that my skin would heal better. She told
me that it was made at one factory in America, and one in Ireland, from where I would get such a suit, tailored exactly to
my size. She told me I would need to wear trousers, a shirt,
gloves, and a mask on my face. Since the suit fit exactly, they
would press against my skin all the time, and when I moved,
the Jobst suit would slightly massage my skin, causing the red-
ness and the hypergrowth of the scars to decrease.

How excited I was! Shula, the physiotherapist, would tell
me about how wonderful the Jobst was. She told me that it
was made in different colors, and immediately I imagined myself covered from head to toe in a tight blue skin, like Spider-
Man; but Shula cautioned me that the colors were only brown
for white people and black for black people. She told me that
people used to call the police when a person wearing the Jobst
mask went into a bank, because they thought it was a bank
robber. Now when you get the mask from the factory, there is
a sign you have to put on your chest, explaining the situation.

Rather than deterring me, this new information made the
suit seem even better. It made me smile. I thought it would be
nice to walk in the streets and actually be invisible. No one
would be able to see any part of me except my mouth and my
eyes. And no one would be able to see my scars.

As I imagined this silky cover, I felt I could endure any
pain until my Jobst suit arrived. Weeks went by. And then it
did arrive. Shula came to help me put it on for the first time.
We started with the trousers: She opened them, in all their
brownish glory, and started to put them on my legs. The feel-

ing wasn't silky like something that would gently massage my scars. The material felt more like canvas that would tear my scars. I was still by no means disillusioned. I wanted to feel how it would be to be immersed completely in the suit.

After a few minutes it became apparent that I had gained some weight since the time when the measurements were taken (they used to feed me 7,000 calories and 30 eggs a day to help my body heal). The Jobst suit didn't fit very well. Still, I had waited a long time for it. Finally, with some stretching and a lot of patience on everyone's part, I was eventually completely dressed. The shirt with the long sleeves put great pressure on my chest, shoulders, and arms. The mask pressed hard all the time. The long trousers began at my toes and went all the way up to my belly button. And there were the gloves. The only visible parts of me were the ends of my toes, my eyes, my ears, and my mouth. Everything else was covered by the brown Jobst.

The pressure seemed to become stronger every minute. The heat inside was intense. My scars had a poor blood supply, and the heat made the blood rush to them, making them red and much more itchy. Even the sign warning people that I was not a bank robber was a failure. The sign was in English, not Hebrew, and so was quite worthless. My lovely dream had failed me. I struggled out of the suit. New measurements were taken and sent to Ireland so that I could get a better-fitting Jobst.

The next suit provided a more comfortable fit, but otherwise it was not much better. I suffered with this treatment for months—itching, aching, struggling to wear it, and tearing my delicate new skin while trying to put it on (and when this new thin skin tears, it takes a long while to heal). At the end I learned that this suit had no real benefits, at least not for me. The areas of my body that were better covered looked and felt no different from the areas that were not as well covered, and the suffering that went along with the suit turned out to be all that it provided me.

You see, while it would be morally questionable to make patients in the burn department take part in an experiment that was designed to test the efficacy of such suits (using different types of fabrics, different pressure levels, etc.), and even more difficult to ask someone to participate in a placebo experiment, it is also morally difficult to inflict painful treatments on many patients and for many years, without having a really good reason to do so.

If this type of synthetic suit had been tested relative to other methods, and relative to a placebo suit, that approach might have eliminated part of my daily misery. It might also have stimulated research on new approaches—ones that would actually work. My wasted suffering, and the suffering of other patients like me, is the real cost of not doing such experiments.

Should we always test every procedure and carry out placebo experiments? The moral dilemmas involved in medical and placebo experiments are real. The potential benefits of such experiments should be weighed against their costs, and as a consequence we cannot, and should not, always do placebo tests. But my feeling is that we are not doing nearly as many of them as we should.

The Context of Our Character, Part I

Why We Are Dishonest, and What We Can Do about It

In 2004, the total cost of all robberies in the United States was $525 million, and the average loss from a single robbery was about $1,300.[18] These amounts are not very high, when we consider how much police, judicial, and corrections muscle is put into the capture and confinement of robbers—let alone the amount of newspaper and television coverage these kinds of crimes elicit. I'm not suggesting that we go easy on career criminals, of course. They are thieves, and we must protect ourselves from their acts.

But consider this: every year, employees' theft and fraud at the workplace are estimated at about $600 billion. That figure is dramatically higher than the combined financial cost of robbery, burglary, larceny-theft, and automobile theft (totaling about $16 billion in 2004); it is much more than what all the career criminals in the United States could steal in their lifetimes; and it's also almost twice the market capitalization of General Electric. But there's much more. Each year, according to reports by the insurance industry, individuals add a bogus $24 billion to their claims of property losses. The IRS, meanwhile, estimates a loss of $350 billion per year, representing the gap between what the feds think people should pay in taxes and what they do pay. The retail industry has its own headache: it loses $16 billion a year to customers who buy clothes, wear them with the tags tucked in, and return these secondhand clothes for a full refund.

Add to this sundry everyday examples of dishonesty—the congressman accepting golfing junkets from his favorite lobbyist; the physician making kickback deals with the laboratories that he uses; the corporate executive who backdates his stock options to boost his final pay—and you have a huge amount of unsavory economic activity, dramatically larger than that of the standard household crooks.

When the Enron scandal erupted in 2001 (and it became apparent that Enron, as *Fortune* magazine's "America's Most Innovative Company" for six consecutive years, owed much of its success to innovations in accounting), Nina Mazar, On Amir (a professor at the University of California at San Diego), and I found ourselves discussing the subject of dishonesty over lunch. Why are some crimes, particularly white-collar crimes, judged less severely than others, we wondered—especially since their perpetrators can inflict more financial damage between their ten o'clock latte and lunch than a standard-issue burglar might in a lifetime?

After some discussion we decided that there might be two types of dishonesty. One is the type of dishonesty that evokes the image of a pair of crooks circling a gas station. As they cruise by, they consider how much money is in the till, who might be around to stop them, and what punishment they may face if caught (including how much time off they might get for good behavior). On the basis of this cost-benefit calculation, they decide whether to rob the place or not.

Then there is the second type of dishonesty. This is the kind committed by people who generally consider themselves honest—the men and women (please stand) who have "borrowed" a pen from a conference site, taken an extra splash of soda from the soft drink dispenser, exaggerated the cost of their television on their property loss report, or falsely reported a meal with Aunt Enid as a business expense (well, she did inquire about how work was going).

We know that this second kind of dishonesty exists, but how prevalent is it? Furthermore, if we put a group of "honest" people into a scientifically controlled experiment and tempted them to cheat, would they? Would they compromise their integrity? Just how much would they steal? We decided to find out.

THE HARVARD BUSINESS SCHOOL holds a place of distinction in American life. Set on the banks of the River Charles in Boston, Massachusetts; housed in imposing colonial-style architecture; and dripping with endowment money, the school is famous for creating America's top business leaders. In the Fortune 500 companies, in fact, about 20 percent of the top three positions are held by graduates of the Harvard Business School.* What better place, then, to do a little experiment on the issue of honesty?†

---

*As claimed by the Harvard Business School.

† We often conduct our experiments at Harvard, not because we think its students are different from MIT's students, but because it has wonderful facilities and the faculty members are very generous in letting us use them.

The study would be fairly simple. We would ask a group of Harvard undergraduates and MBA students to take a test consisting of 50 multiple-choice questions. The questions would be similar to those on standardized tests (What is the longest river in the world? Who wrote *Moby-Dick*? What word describes the average of a series? Who, in Greek mythology, was the goddess of love?). The students would have 15 minutes to answer the questions. At the end of that time, they would be asked to transfer their answers from their worksheet to a scoring sheet (called a bubble sheet), and submit both the worksheet and the bubble sheet to a proctor at the front of the room. For every correct answer, the proctor would hand them 10 cents. Simple enough.

In another setup we asked a new group of students to take the same general test, but with one important change. The students in this section would take the test and transfer their work to their scoring bubble sheet, as the previous group did. But this time the bubble sheet would have the correct answers pre-marked. For each question, the bubble indicating the correct answer was colored gray. If the students indicated on their worksheet that the longest river in the world is the Mississippi, for instance, once they received the bubble sheet, they would clearly see from the markings that the right answer is the Nile. At that point, if the participants chose the wrong answer on their worksheet, they could decide to lie and mark the correct answer on the bubble sheet.

After they transferred their answers, they counted how many questions they had answered correctly, wrote that number at the top of their bubble sheet, and handed both the worksheet and the bubble sheet to the proctor at the front of the room. The proctor looked at the number of questions they claimed to have answered correctly (the summary number they wrote at the top of the bubble sheet) and paid them 10 cents per correct answer.

Would the students cheat—changing their wrong answers to the ones pre-marked on the bubble sheet? We weren't sure, but in any case, we decided to tempt the next group of students even more. In this condition the students would again take the test and transfer their answers to the pre-marked bubble sheet. But this time we would instruct them to shred their original worksheet, and hand only the bubble sheet to the proctor. In other words, they would destroy all evidence of any possible malfeasance. Would they take the bait? Again, we didn't know.

In the final condition, we would push the group's integrity to the limit. This time they would be instructed to destroy not only their original worksheet, but the final pre-marked bubble sheet as well. Moreover, they wouldn't even have to report their earnings to the experimenter: When they were finished shredding their work and answer sheets, they merely needed to walk up to the front of the room—where we had placed a jar full of coins—withdraw their earnings, and saunter out the door. If one was ever inclined to cheat, this was the opportunity to pull off the perfect crime.

Yes, we were tempting them. We were making it easy to cheat. Would the crème de la crème of America's youth take the bait? We'd have to see.

As the first group settled into their seats, we explained the rules and handed out the tests. They worked for their 15 minutes, then copied their answers onto the bubble sheet, and turned in their worksheets and bubble sheets. These students were our control group. Since they hadn't been given any of the answers, they had no opportunity at all to cheat. On average, they got 32.6 of the 50 questions right.

What do you predict that the participants in our other experimental conditions did? Given that the participants in the control condition solved on average 32.6 questions correctly, how many questions do you think the participants in the other three conditions claimed to have solved correctly?

Condition 1	Control	=	32.6
Condition 2	Self-check	=	_____
Condition 3	Self-check + shredding	=	_____
Condition 4	Self-check + shredding +money jar	=	_____

What about the second group? They too answered the questions. But this time, when they transferred their answers to the bubble sheet, they could see the correct answers. Would they sweep their integrity under the rug for an extra 10 cents per question? As it turned out, this group claimed to have solved on average 36.2 questions. Were they smarter than our control group? Doubtful. Instead, we had caught them in a bit of cheating (by about 3.6 questions).

What about the third group? This time we upped the ante. They not only got to see the correct answers but were also asked to shred their worksheets. Did they take the bait? Yes, they cheated. On average they claimed to have solved 35.9 questions correctly—more than the participants in the control condition, but about the same as the participants in the second group (the group that did not shred their worksheets).

Finally came the students who were told to shred not only their worksheets but the bubble sheets as well—and then dip their hands into the money jar and withdraw whatever they deserved. Like angels they shredded their worksheets, stuck their hands into the money jar, and withdrew their coins. The problem was that these angels had dirty faces: their claims added up to an average 36.1 correct answers—quite a bit higher than the 32.6 of our control group, but basically the same as the other two groups who had the opportunity to cheat.

What did we learn from this experiment? The first conclusion, is that when given the opportunity, many honest people will cheat. In fact, rather than finding that a few bad apples weighted the averages, we discovered that the majority of people cheated, and that they cheated just a little bit.* And before you blame the refined air at the Harvard Business School for this level of dishonesty, I should add that we conducted similar experiments at MIT, Princeton, UCLA, and Yale with similar results.

The second, and more counterintuitive, result was even more impressive: once tempted to cheat, the participants didn't seem to be as influenced by the risk of being caught as one might think. When the students were given the opportunity to cheat without being able to shred their papers, they increased their correct answers from 32.6 to 36.2. But when they were offered the chance to shred their papers—hiding their little crime completely—they didn't push their dishonesty farther. They still cheated at about the same level. This means that even when we have no chance of getting caught, we still don't become wildly dishonest.

---

*The distribution of the number of correctly solved questions remained the same across all four conditions, but with a mean shift when the participants could cheat.

When the students could shred both their papers, dip their hand into the money jar, and walk away, every one of them could have claimed a perfect test score, or could have taken more money (the jar had about $100 in it). But none of them did. Why? Something held them back—something inside them. But what was it? What is honesty, anyhow?

To that question, Adam Smith, the great economic thinker, had a pleasant reply: "Nature, when she formed man for society, endowed him with an original desire to please, and an original aversion to offend his bretheren. She taught him to feel pleasure in their favourable, and pain in their unfavourable regard," he noted.

To this Smith added, "The success of most people . . . almost always depends upon the favour and good opinion of their neighbours and equals; and without a tolerably regular conduct these can very seldom be obtained. The good old proverb, therefore, that honesty is always the best policy, holds, in such situations, almost always perfectly true."

That sounds like a plausible industrial-age explanation, as balanced and harmonious as a set of balance weights and perfectly meshed gears. However optimistic this perspective might seem, Smith's theory had a darker corollary: since people engage in a cost-benefit analysis with regard to honesty, they can also engage in a cost-benefit analysis to be dishonest. According to this perspective, individuals are honest only to the extent that suits them (including their desire to please others).

Are decisions about honesty and dishonesty based on the same cost-benefit analysis that we use to decide between cars, cheeses, and computers? I don't think so. First of all, can you imagine a friend explaining to you the cost-benefit analysis that went into buying his new laptop? Of course. But can you imagine your friend sharing with you a cost-benefit analysis of her decision to steal a laptop? Of course not—not unless your friend is a professional thief. Rather, I agree with others (from Plato down) who say that honesty is something bigger—something that is considered a moral virtue in nearly every society.

Sigmund Freud explained it this way. He said that as we grow up in society, we internalize the social virtues. This internalization leads to the development of the superego. In general, the superego is pleased when we comply with society's ethics, and unhappy when we don't. This is why we stop our car at four AM when we see a red light, even if we know that no one is around; and it is why we get a warm feeling when we return a lost wallet to its owner, even if our identity is never revealed. Such acts stimulate the reward centers of our brain—the nucleus accumbens and the caudate nucleus—and make us content.

But if honesty is important to us (in a recent survey of nearly 36,000 high school students in the United States, 98 percent of them said it was important to be honest), and if honesty makes us feel good, why are we so frequently dishonest?

This is my take. We care about honesty and we want to be honest. The problem is that our internal honesty monitor is active only when we contemplate big transgressions, like grabbing an entire box of pens from the conference hall. For the little transgressions, like taking a single pen or two pens, we don't even consider how these actions would reflect on our honesty and so our superego stays asleep.

Without the superego's help, monitoring, and managing

of our honesty, the only defense we have against this kind of transgression is a rational cost-benefit analysis. But who is going to consciously weigh the benefits of taking a towel from a hotel room versus the cost of being caught? Who is going to consider the costs and benefits of adding a few receipts to a tax statement? As we saw in the experiment at Harvard, the cost-benefit analysis, and the probability of getting caught in particular, does not seem to have much influence on dishonesty.

THIS IS THE way the world turns. It's almost impossible to open a newspaper without seeing a report of a dishonest or deceptive act. We watch as the credit card companies bleed their customers with outrageous interest rate hikes; as the airlines plunge into bankruptcy and then call on the federal government to get them—and their underfunded pension funds—out of trouble; and as schools defend the presence of soda machines on campus (and rake in millions from the soft drinks firms) all the while knowing that sugary drinks make kids hyperactive and fat. Taxes are a festival of eroding ethics, as the insightful and talented reporter David Cay Johnston of the *New York Times* describes in his book *Perfectly Legal: The Covert Campaign to Rig Our Tax System to Benefit the Super Rich—and Cheat Everybody Else.*

Against all of this, society, in the form of the government, has battled back, at least to some extent. The Sarbanes-Oxley Act of 2002 (which requires the chief executives of public companies to vouch for the firms' audits and accounts) was passed to make debacles like Enron's a thing of the past. Congress has also passed restrictions on "earmarking" (specifically the pork-barrel spending that politicians insert into larger federal bills). The Securities and Exchange Commission even passed requirements for additional disclosure about executives' pay and perks—so that when we see a stretch limo carrying a Fortune 500 executive, we know pretty certainly how much the corporate chief inside is getting paid.

But can external measures like these really plug all the holes and prevent dishonesty? Some critics say they can't. Take the ethics reforms in Congress, for instance. The statutes ban lobbyists from serving free meals to congressmen and their aides at "widely attended" functions. So what have lobbyists done? Invited congressmen to luncheons with "limited" guest lists that circumvent the rule. Similarly, the new ethics laws ban lobbyists from flying congressmen in "fixed-wing" aircraft. So hey, how about a lift in a helicopter?

The most amusing new law I've heard about is called the "toothpick rule." It states that although lobbyists can no longer provide sit-down meals to congressmen, the lobbyists can still serve anything (presumably hors d'oeuvres) which the legislators can eat while standing up, plopping into their mouths using their fingers or a toothpick.

Did this change the plans of the seafood industry, which had organized a sit-down pasta and oyster dinner for Washington's legislators (called—you guessed it—"Let the World Be Your Oyster")? Not by much. The seafood lobbyists did drop the pasta dish (too messy to fork up with a toothpick), but still fed the congressmen well with freshly opened raw oysters (which the congressmen slurped down standing up).[19]

Sarbanes-Oxley has also been called ineffectual. Some critics say that it's rigid and inflexible, but the loudest outcry is from those who call it ambiguous, inconsistent, inefficient, and outrageously expensive (especially for smaller firms). "It hasn't cleaned up corruption," argued William A. Niskanen, chairman of the Cato Institute; "it has only forced companies to jump through hoops."

So much for enforcing honesty through external controls. They may work in some cases, but not in others. Could there be a better cure for dishonesty?

BEFORE I EVEN attempt to answer that question, let me describe an experiment we conducted that speaks volumes on the subject. A few years ago Nina, On, and I brought a group of participants together in a lab at UCLA and asked them to take a simple math test. The test consisted of 20 simple problems, each requiring participants to find two numbers that would add up to 10 (for a sample problem, see the table below). They had five minutes to solve as many of the problems as they could, after which they were entered into a lottery. If they won the lottery, they would receive ten dollars for each problem they solved correctly.

As in our experiment at the Harvard Business School, some of the participants handed in their papers directly to the experimenter. They were our control group. The other participants wrote down on another sheet the number of questions they solved correctly, and then disposed of the originals. These participants, obviously, were the ones with

Look at your watch, note the time, and start searching for two numbers in the matrix below that will add up to exactly 10. How long did it take you?

1.69	1.82	2.91
4.67	4.81	3.05
5.82	5.06	4.28
6.36	5.19	4.57

the opportunity to cheat. So, given this opportunity, did these participants cheat? As you may have surmised, they did (but, of course, just by a bit).

Up to now I have not told you anything new. But the key to this experiment was what preceded it. When the participants first came to the lab, we asked some of them to write down the names of 10 books that they read in high school. The others were asked to write down as many of the Ten Commandments as they could recall.* After they finished this "memory" part of the experiment, we asked them to begin working on the matrix task.

This experimental setup meant that some of the participants were tempted to cheat after recalling 10 books that they read in high school, and some of them were tempted after recalling the Ten Commandments. Who do you think cheated more?

When cheating was *not* possible, our participants, on average, solved 3.1 problems correctly.†

When cheating was possible, the group that recalled 10 books read in high school achieved an average score of 4.1 questions solved (or 33 percent more than those who could not cheat).

But the big question is what happened to the other group—the students who first wrote down the Ten Commandments, then took the test, and then ripped up their worksheets. This, as sportscasters say, was the group to watch. Would they cheat—or would the Ten Commandments have an effect on

---

*Do you know the Ten Commandments? If you'd like to test yourself, write them down and compare your list with the list at the end of this chapter. To be sure you have them right, don't just say them to yourself; write them down.
†Can the Ten Commandments raise one's math scores? We used the same two memory tasks with the control condition to test that premise. The performance in the control condition was the same regardless of the type of memory task. So the Commandments do not raise math scores.

their integrity? The result surprised even us: the students who had been asked to recall the Ten Commandments had not cheated at all. They averaged three correct answers—the same basic score as the group that could not cheat, and one less than those who were able to cheat but had recalled the names of the books.

As I walked home that evening I began to think about what had just happened. The group who listed 10 books cheated. Not a lot, certainly—only to that point where their internal reward mechanism (nucleus accumbens and super-ego) kicked in and rewarded them for stopping.

But what a miracle the Ten Commandments had wrought! We didn't even remind our participants what the Command-ments were—we just asked each participant to recall them (and almost none of the participants could recall all 10). We hoped the exercise might evoke the idea of honesty among them. And this was clearly what it did. So, we wondered, what lessons about decreasing dishonesty can we learn from this experi-ment? It took us a few weeks to come to some conclusions.

FOR ONE, PERHAPS we could bring the Bible back into public life. If we only want to reduce dishonesty, it might not be a bad idea. Then again, some people might object, on the grounds that the Bible implies an endorsement of a particular religion, or merely that it mixes religion in with the commer-cial and secular world. But perhaps an oath of a different nature would work. What especially impressed me about the experiment with the Ten Commandments was that the stu-dents who could remember only one or two Commandments were as affected by them as the students who remembered nearly all ten. This indicated that it was not the Commandments themselves that encouraged honesty, but the mere contemplation of a moral benchmark of some kind.

If that were the case, then we could also use nonreligious benchmarks to raise the general level of honesty. For instance, what about the professional oaths that doctors, lawyers, and others swear to—or used to swear to? Could those professional oaths do the trick?

The word *profession* comes from the Latin *professus*, meaning "affirmed publicly." Professions started somewhere deep in the past in religion and then spread to medicine and law. Individuals who had mastered esoteric knowledge, it was said, not only had a monopoly on the practice of that knowledge, but had an obligation to use their power wisely and honestly. The oath—spoken and often written—was a reminder to practitioners to regulate their own behavior, and it also provided a set of rules that had to be followed in fulfilling the duties of their profession.

Those oaths lasted a long time. But then, in the 1960s, a strong movement arose to deregulate professions. Professions were elitist organizations, it was argued, and needed to be turned out into the light of day. For the legal profession that meant more briefs written in plain English prose, cameras in the courtrooms, and advertising. Similar measures against elitism were applied to medicine, banking, and other professions as well. Much of this could have been beneficial, but something was lost when professions were dismantled. Strict professionalism was replaced by flexibility, individual judgment, the laws of commerce, and the urge for wealth, and with it disappeared the bedrock of ethics and values on which the professions had been built.

A study by the state bar of California in the 1990s, for instance, found that a preponderance of attorneys in California were sick of the decline in honor in their work and "profoundly pessimistic" about the condition of the legal profession. Two-thirds said that lawyers today "compromise their professionalism as a result of economic pressure." Nearly 80 percent said that the bar "fails to adequately punish unethical attorneys." Half said they wouldn't become attorneys if they had it to do over again.[20]

A comparable study by the Maryland Judicial Task Force found similar distress among lawyers in that state. According to Maryland's lawyers, their profession had degenerated so badly that "they were often irritable, short-tempered, argumentative, and verbally abusive" or "detached, withdrawn, preoccupied, or distracted." When lawyers in Virginia were asked whether the increasing problems with professionalism were attributable to "a few bad apples" or to a widespread trend, they overwhelmingly said this was a widespread issue.[21]

Lawyers in Florida have been deemed the worst.[22] In 2003 the Florida bar reported that a "substantial minority" of lawyers were "money-grabbing, too clever, tricky, sneaky, and not trustworthy; who had little regard for the truth or fairness, willing to distort, manipulate, and conceal to win; arrogant, condescending, and abusive." They were also "pompous and obnoxious." What more can I say?

The medical profession has its critics as well. The critics mention doctors who do unnecessary surgeries and other procedures just to boost the bottom line: who order tests at laboratories that are giving them kickbacks, and who lean toward medical tests on equipment that they just happen to own. And what about the influence of the pharmaceutical industry? A friend of mine said he sat waiting for his doctor for an hour recently. During that time, he said, four (very attractive) representatives of drug companies went freely into and out of the office, bringing lunch, free samples, and other gifts with them.

You could look at almost any professional group and see signs of similar problems. How about the Association of Petroleum Geologists, for instance? The image I see is Indiana Jones types, with more interest in discussing Jurassic shale and deltaic deposits than in making a buck. But look deeper and you'll find trouble. "There is unethical behavior going on at a much larger scale than most of us would care to think," one member of the association wrote to her colleagues.[23]

What kind of dishonesty, for goodness' sake, could be rife in the ranks of petroleum geologists, you ask? Apparently things like using bootlegged seismic and digital data; stealing maps and materials; and exaggerating the promise of certain oil deposits, in cases where a land sale or investment is being made. "The malfeasance is most frequently of shades of gray, rather than black and white," one petroleum geologist remarked.

But let's remember that petroleum geologists are not alone. This decline in professionalism is everywhere. If you need more proof, consider the debate within the field of professional ethicists, who are called more often than ever before to testify at public hearings and trials, where they may be hired by one party or another to consider issues such as treatment rendered to a patient and the rights of the unborn. Are they tempted to bend to the occasion? Apparently so. "Moral Expertise: A Problem in the Professional Ethics of Professional Ethicists" is the title of one article in an ethics journal.[24] As I said, the signs of erosion are everywhere.

WHAT TO DO? Suppose that, rather than invoking the Ten Commandments, we got into the habit of signing our name to some secular statement—similar to a professional oath—that would remind us of our commitment to honesty. Would a simple oath make a difference, in the way that we saw the Ten Commandments make a difference? We needed to find out—hence our next experiment.

Once again we assembled our participants. In this study, the first group of participants took our matrix math test and handed in their answers to the experimenter in the front of the room (who counted how many questions they answered correctly and paid them accordingly). The second group also took the test, but the members of this group were told to fold their answer sheet, keep it in their possession, and tell the experimenter in the front of the room how many of the problems they got right. The experimenter paid them accordingly, and they were on their way.

The novel aspect of this experiment had to do with the third group. Before these participants began, each was asked to sign the following statement on the answer sheet: "I understand that this study falls under the MIT honor system." After signing this statement, they continued with the task. When the time had elapsed they pocketed their answer sheets, walked to the front of the room, told the experimenter how many problems they had correctly solved, and were paid accordingly.

What were the results? In the control condition, in which cheating was not possible, participants solved on average three problems (out of 20). In the second condition, in which the participants could pocket their answers, they claimed to have solved on average 5.5 problems. What was remarkable was the third situation—in which the participants pocketed their answer sheets, but had also signed the honor code statement. In this case they claimed to have solved, on average, three problems—exactly the same number as the control group. This outcome was similar to the results we achieved with the Ten Commandments—when a moral reminder eliminated cheating altogether. The effect of signing a statement about an honor code is particularly amazing when we take into account that MIT doesn't even have an honor code.

So we learned that people cheat when they have a chance to do so, but they don't cheat as much as they could. Moreover, once they begin thinking about honesty—whether by recalling the Ten Commandments or by signing a simple statement—they stop cheating completely. In other words, when we are removed from any benchmarks of ethical thought, we tend to stray into dishonesty. But if we are reminded of morality at the moment we are tempted, then we are much more likely to be honest.

At present, several state bars and professional organizations are scrambling to shore up their professional ethics. Some are increasing courses in college and graduate schools, and others are requiring brush-up ethics classes. In the legal profession, Judge Dennis M. Sweeney of the Howard County (Maryland) circuit published his own book, *Guidelines for Lawyer Courtroom Conduct*, in which he noted, "Most rules, like these, are simply what our mothers would say a polite and well raised man or woman should do. Since, given their other important responsibilities, our mothers (and yours) cannot be in every courtroom in the State, I offer these rules."

Will such general measures work? Let's remember that lawyers do take an oath when they are admitted to the bar, as doctors take an oath when they enter their profession. But occasional swearing of oaths and occasional statements of adherence to rules are not enough. From our experiments, it is clear that oaths and rules must be recalled at, or just before, the moment of temptation. Also, what is more, time is working against us as we try to curb this problem. I said in Chapter 4 that when social norms collide with market norms, the social norms go away and the market norms stay. Even if the analogy is not exact, honesty offers a related lesson: once professional ethics (the social norms) have declined, getting them back won't be easy.

THIS DOESN'T MEAN that we shouldn't try. Why is honesty so important? For one thing, let's not forget that the United States holds a position of economic power in the world today partly because it is (or at least is perceived to be) one of the world's most honest nations, in terms of its standards of corporate governance.

In 2002, the United States ranked twentieth in the world in terms of integrity, according to one survey (Denmark, Finland, and New Zealand were first; Haiti, Iraq, Myanmar, and Somalia were last, at number 163). On this basis, I would suspect that people doing business with the United States generally feel they can get a fair deal. But the fact of the matter is that the United States ranked fourteenth in 2000, before the wave of corporate scandals made the business pages in American newspapers look like a police blotter.[25] We are going down the slippery slope, in other words, not up it, and this can have tremendous long-term costs.

Adam Smith reminded us that honesty really is the best policy, especially in business. To get a glimpse at the other side of that realization—at the downside, in a society without trust—you can take a look at several countries. In China, the word of one person in one region rarely carries to another region. Latin America is full of family-run cartels that hand out loans to relatives (and then fail to cut off credit when the debtor begins to default). Iran is another example of a nation stricken by distrust. An Iranian student at MIT told me that business there lacks a platform of trust. Because of this, no one pays in advance, no one offers credit, and no one is willing to take risks. People must hire within their families, where some level of trust still exists. Would you like to live in such a world? Be careful, because without honesty we might get there faster than you'd imagine.

What can we do to keep our country honest? We can read the Bible, the Koran, or whatever reflects our values, perhaps. We can revive professional standards. We can sign our names to promises that we will act with integrity. Another path is to first recognize that when we get into situations where our personal financial benefit stands in opposition to our moral standards, we are able to "bend" reality, see the world in terms compatible with our selfish interest, and become dishonest. What is the answer, then? If we recognize this weakness, we can try to avoid such situations from the outset. We can prohibit physicians from ordering tests that would benefit them financially; we can prohibit accountants and auditors from functioning as consultants to the same companies; we can bar members of Congress from setting their own salaries, and so on.

But this is not the end of the issue of dishonesty. In the next chapter, I will offer some other suggestions about dishonesty, and some other insights into how we struggle with it.

APPENDIX: CHAPTER 11
The Ten Commandments[*]

I am the Lord your God, you shall have
no other gods before me.

You shall not take the name of the Lord your God in vain.

Keep holy the Sabbath day.

Honor your father and your mother.

You shall not kill.

You shall not commit adultery.

You shall not steal.

You shall not bear false witness.

You shall not covet your neighbor's wife.

You shall not covet your neighbor's goods.

[*]There are several versions of the Ten Commandments. I randomly chose this Roman Catholic version.

The Context of Our Character, Part II

Why Dealing with Cash Makes Us More Honest

M any of the dormitories at MIT have common areas, where sit a variety of refrigerators that can be used by the students in the nearby rooms. One morning at about eleven, when most of the students were in class, I slipped into the dorms and, floor by floor, went hunting for all the shared refrigerators that I could find.

When I detected a communal fridge, I inched toward it. Glancing cautiously around, I opened the door, slipped in a six-pack of Coke, and walked briskly away. At a safe distance, I paused and jotted down the time and the location of the fridge where I had left my Cokes.

Over the next few days I returned to check on my Coke cans. I kept a diary detailing how many of them remained in the fridge. As you might expect, the half-life of Coke in a college dorm isn't very long. All of them had vanished within 72 hours. But I didn't always leave Cokes behind. In some of the fridges, I left a plate containing six one-dollar bills. Would the money disappear faster than the Cokes?

Before I answer that question, let me ask you one. Suppose your spouse calls you at work. Your daughter needs a red pencil for school the next day. "Could you bring one home?" How comfortable would you be taking a red pencil from work for your daughter? Very uncomfortable? Somewhat uncomfortable? Completely comfortable?

Let me ask you another question. Suppose there are no red pencils at work, but you can buy one downstairs for a dime. And the petty cash box in your office has been left open, and no one is around. Would you take 10 cents from the petty cash box to buy the red pencil? Suppose you didn't have any change and needed the 10 cents. Would you feel comfortable taking it? Would that be OK?

I don't know about you, but while I'd find taking a red pencil from work relatively easy, I'd have a very hard time taking the cash. (Luckily for me, I haven't had to face this issue, since my daughter is not in school yet.)

As it turns out, the students at MIT also felt differently about taking cash. As I mentioned, the cans of Coke quickly disappeared; within 72 hours every one of them was gone. But what a different story with the money! The plates of dollar bills remained untouched for 72 hours, until I removed them from the refrigerators.

So what's going on here?

When we look at the world around us, much of the dishonesty we see involves cheating that is one step removed from cash. Companies cheat with their accounting practices; executives cheat by using backdated stock options; lobbyists cheat by underwriting parties for politicians; drug companies cheat by sending doctors and their wives off on posh vacations. To be sure, these people don't cheat with cold cash (except occasionally). And that's my point: cheating is a lot easier when it's a step removed from money.

Do you think that the architects of Enron's collapse—Kenneth Lay, Jeffrey Skilling, and Andrew Fastow—would have stolen money from the purses of old women? Certainly, they took millions of dollars in pension monies from a lot of old women. But do you think they would have hit a woman with a blackjack and pulled the cash from her fingers? You may disagree, but my inclination is to say no.

So what permits us to cheat when cheating involves non-monetary objects, and what restrains us when we are dealing with money? How does that irrational impulse work?

BECAUSE WE ARE so adept at rationalizing our petty dishonesty, it's often hard to get a clear picture of how nonmonetary objects influence our cheating. In taking a pencil, for example, we might reason that office supplies are part of our overall compensation, or that lifting a pencil or two is what everyone does. We might say that taking a can of Coke from a communal refrigerator from time to time is all right, because, after all, we've all had cans of Coke taken from us. Maybe Lay, Skilling, and Fastow thought that cooking the books at Enron was OK, since it was a temporary measure that could be corrected when business improved. Who knows?

To get at the true nature of dishonesty, then, we needed to develop a clever experiment, one in which the object in question would allow few excuses. Nina, On, and I thought about it. Suppose we used symbolic currency, such as tokens. They were not cash, but neither were they objects with a history, like a Coke or a pencil. Would it give us insight into the cheating process? We weren't sure, but it seemed reasonable; and so, a few years ago, we gave it a try.

This is what happened. As the students at one of the MIT cafeterias finished their lunches, we interrupted them to ask whether they would like to participate in a five-minute experiment. All they had to do, we explained, was solve 20 simple math problems (finding two numbers that added up to 10). And for this they would get 50 cents per correct answer.

The experiment began similarly in each case, but ended in one of three different ways. When the participants in the first group finished their tests, they took their worksheets up to the experimenter, who tallied their correct answers and paid them 50 cents for each. The participants in the second group were told to tear up their worksheets, stuff the scraps into their pockets or backpacks, and simply tell the experimenter their score in exchange for payment. So far this experiment was similar to the tests of honesty described in the previous chapter.

But the participants in the last group had something significantly different in their instructions. We told them, as we had told the previous group, to tear up the worksheets and simply tell the experimenter how many questions they had answered correctly. But this time, the experimenter wouldn't be giving them cash. Rather, she would give them a token for each question they claimed to have solved. The students would then walk 12 feet across the room to another experimenter, who would exchange each token for 50 cents.

Do you see what we were doing? Would the insertion of a token into the transaction—a piece of valueless, nonmonetary currency—affect the students' honesty? Would the token make the students less honest in tallying their answers than the students who received cash immediately? If so, by how much?

Even we were surprised by the results: The participants in the first group (who had no way to cheat) solved an average of 3.5 questions correctly (they were our control group).

The participants in the second group, who tore up their worksheets, claimed to have correctly solved an average of 6.2 questions. Since we can assume that these students did not become smarter merely by tearing up their worksheets, we can attribute the 2.7 additional questions they claimed to have solved to cheating.

But in terms of brazen dishonesty, the participants in the third group took the cake. They were no smarter than the previous two groups, but they claimed to have solved an average of 9.4 problems—5.9 more than the control group and 3.2 more than the group that merely ripped up the worksheets.

This means that when given a chance to cheat under ordinary circumstances, the students cheated, on average, by 2.7 questions. But when they were given the same chance to cheat with nonmonetary currency, their cheating increased to 5.9—more than doubling in magnitude. What a difference there is in cheating for money versus cheating for something that is a step away from cash!

If that surprises you, consider this. Of the 2,000 participants in our studies of honesty (described in the previous chapter), only four ever claimed to have solved all the problems. In other words, the rate of "total cheating" was four in 2,000.*

---

*Theoretically, it is possible that some people solved all the problems. But since no one in the control conditions solved more than 10 problems, the likelihood that four of our participants truly solved 20 is very, very low. For this reason we assumed that they cheated.

But in the experiment in which we inserted nonmonetary currency (the token), 24 of the study's 450 participants cheated "all the way." How many of these 24 extreme cheaters were in the condition with money versus the condition with tokens? They were all in the token condition (24 of 150 students cheated "all the way" in this condition; this is equivalent to about 320 per 2,000 participants). This means that not only did the tokens "release" people from some of their moral constraints, but for quite a few of them, the extent of the release was so complete that they cheated as much as was possible.

This level of cheating is clearly bad, but it could have been worse. Let's not forget that the tokens in our experiments were transformed into cash within a matter of seconds. What would the rate of dishonesty have been if the transfer from a nonmonetary token to cash took a few days, weeks, or months (as, for instance, in a stock option)? Would even more people cheat, and to a larger extent?

WE HAVE LEARNED that given a chance, people cheat. But what's really odd is that most of us don't see this coming. When we asked students in another experiment to predict if people would cheat more for tokens than for cash, the students said no, the amount of cheating would be the same. After all, they explained, the tokens represented real money— and the tokens were exchanged within seconds for actual cash. And so, they predicted, our participants would treat the tokens as real cash.

But how wrong they were! They didn't see how fast we can rationalize our dishonesty when it is one step away from cash. Of course, their blindness is ours as well. Perhaps it's why so much cheating goes on. Perhaps it's why Jeff Skilling, Bernie Ebbers, and the entire roster of executives who have been prosecuted in recent years let themselves, and their companies, slide down the slope.

All of us are vulnerable to this weakness, of course. Think about all the insurance fraud that goes on. It is estimated that when consumers report losses on their homes and cars, they creatively stretch their claims by about 10 percent. (Of course, as soon as you report an exaggerated loss, the insurance company raises its rates, so the situation becomes tit for tat). Again it is not the case that there are many claims that are completely flagrant, but instead many people who have lost, say, a 27-inch television set report the loss of a 32-inch set; those who have lost a 32-inch set report the loss of a 36-inch set, and so on. These same people would be unlikely to steal money directly from the insurance companies (as tempting as that might sometimes be), but reporting what they no longer have—and increasing its size and value by just a little bit— makes the moral burden easier to bear.

There are other interesting practices. Have you ever heard the term "wardrobing"? Wardrobing is buying an item of clothing, wearing it for a while, and then returning it in such a state that the store has to accept it but can no longer resell it. By engaging in wardrobing, consumers are not directly stealing money from the company; instead, it is a dance of buying and returning, with many unclear transactions involved. But there is at least one clear consequence—the clothing industry estimates that its annual losses from wardrobing are about $16 billion (about the same amount as the estimated annual loss from home burglaries and automobile theft combined).

And how about expense reports? When people are on

business trips, they are expected to know what the rules are, but expense reports too are one step, and sometimes even a few steps, removed from cash. In one study, Nina and I found that not all expenses are alike in terms of people's ability to justify them as business expenses. For example, buying a mug for five dollars for an attractive stranger was clearly out of bounds, but buying the same stranger an eight-dollar drink in a bar was very easy to justify. The difference was not the cost of the item, or the fear of getting caught, but people's ability to justify the item to themselves as a legitimate use of their expense account.

A few more investigations into expense accounts turned up similar rationalizations. In one study, we found that when people give receipts to their administrative assistants to submit, they are then one additional step removed from the dishonest act, and hence more likely to slip in questionable receipts. In another study, we found that businesspeople who live in New York are more likely to consider a gift for their kid as a business expense if they purchased it at the San Francisco airport (or someplace else far from home) than if they had purchased it at the New York airport, or on their way home from the airport. None of this makes logical sense, but when the medium of exchange is nonmonetary, our ability to rationalize increases by leaps and bounds.

I HAD MY own experience with dishonesty a few years ago. Someone broke into my Skype account (very cool online telephone software) and charged my PayPal account (an online payment system) a few hundred dollars for the service.

I don't think the person who did this was a hardened criminal. From a criminal's perspective, breaking into my account would most likely be a waste of time and talent because if this person was sufficiently smart to hack into Skype, he could probably have hacked into Amazon, Dell, or maybe even a credit card account, and gotten much more value for his time. Rather, I imagine that this person was a smart kid who had managed to hack into my account and who took advantage of this "free" communication by calling anyone who would talk to him until I managed to regain control of my account. He may have even seen this as a techie challenge—or maybe he is a student to whom I once gave a bad grade and who decided to tweak my nose for it.

Would this kid have taken cash from my wallet, even if he knew for sure that no one would ever catch him? Maybe, but I imagine that the answer is no. Instead, I suspect that there were some aspects of Skype and of how my account was set up that "helped" this person engage in this activity and not feel morally reprehensible: First, he stole calling time, not money. Next, he did not gain anything tangible from the transaction. Third, he stole from Skype rather than directly from me. Fourth, he might have imagined that at the end of the day Skype, not I, would cover the cost. Fifth, the cost of the calls was charged automatically to me via PayPal. So here we had another step in the process—and another level of fuzziness in terms of who would eventually pay for the calls. (Just in case you are wondering, I have since canceled this direct link to PayPal.)

Was this person stealing from me? Sure, but there were so many things that made the theft fuzzy that I really don't think he thought of himself as a dishonest guy. No cash was taken, right? And was anyone really hurt? This kind of thinking is worrisome. If my problem with Skype was indeed due to the nonmonetary nature of the transactions on Skype, this would mean that there is much more at risk here, including a wide range of online services, and perhaps even credit and debit cards. All these electronic transactions, with no physical exchange of money from hand to hand, might make it easier for people to be dishonest—without ever questioning or fully acknowledging the immorality of their actions.

THERE'S ANOTHER, SINISTER impression that I took out of our studies. In our experiments, the participants were smart, caring, honorable individuals, who for the most part had a clear limit to the amount of cheating they would undertake, even with nonmonetary currency like the tokens. For almost all of them, there was a point at which their conscience called for them to stop, and they did. Accordingly, the dishonesty that we saw in our experiments was probably the lower boundary of human dishonesty: the level of dishonesty practiced by individuals who want to be ethical and who want to see themselves as ethical—the so-called good people.

The scary thought is that if we did the experiments with nonmonetary currencies that were not as immediately convertible into money as tokens, or with individuals who cared less about their honesty, or with behavior that was not so publicly observable, we would most likely have found even higher levels of dishonesty. In other words, the level of deception we observed here is probably an underestimation of the level of deception we would find across a variety of circumstances and individuals.

Now suppose that you have a company or a division of a company led by a Gordon Gekko character who declares that "greed is good." And suppose he used nonmonetary means of encouraging dishonesty. Can you see how such a swashbuckler could change the mind-set of people who in principle want to be honest and want to see themselves as honest, but also want to hold on to their jobs and get ahead in the world? It is under just such circumstances that non-monetary currencies can lead us astray. They let us bypass our conscience and freely explore the benefits of dishonesty.

This view of human nature is worrisome. We can hope to surround ourselves with good, moral people, but we have to be realistic. Even good people are not immune to being partially blinded by their own minds. This blindness allows them to take actions that bypass their own moral standards on the road to financial rewards. In essence, motivation can play tricks on us whether or not we are good, moral people.

As the author and journalist Upton Sinclair once noted, "It is difficult to get a man to understand something when his salary depends upon his not understanding it." We can now add the following thought: it is even more difficult to get a man to understand something when he is dealing with non-monetary currencies.

THE PROBLEMS OF dishonesty, by the way, don't apply just to individuals. In recent years we have seen business in general succumb to a lower standard of honesty. I'm not talking about big acts of dishonesty, like those perpetrated by Enron and Worldcom. I mean the small acts of dishonesty that are similar to swiping Cokes out of the refrigerator. There are companies out there, in other words, that aren't stealing cash off our plates, so to speak, but are stealing things one step removed from cash.

There are plenty of examples. Recently, one of my friends, who had carefully saved up his frequent-flyer miles for a vacation, went to the airline who issued all these miles. He was told that all the dates he wanted were blacked out. In other words, although he had saved up 25,000 frequent-flyer miles, he couldn't use them (and he tried many dates). But, the representative said, if he wanted to use 50,000 miles, there might be some seats. She checked. Sure, there were seats everywhere.

To be sure, there was probably some small print in the frequently-flyer brochure explaining that this was OK. But to my friend, the 25,000 miles he had earned represented a lot of money. Let's say it was $450. Would this airline have mugged him for that amount of cash? Would the airline have swiped it from his bank account? No. But because it was one step removed, the airline stole it from him in the form of requiring 25,000 additional miles.

For another example, look at what banks are doing with credit card rates. Consider what is called two-cycle billing. There are several variations of this trick, but the basic idea is that the moment you don't pay your bill in full, the credit issuer will not only charge a high interest rate on new purchases, but will actually reach into the past and charge interest on past purchases as well. When the Senate banking committee looked into this recently, it heard plenty of testimony that certainly made the banks look dishonest. For instance, a man in Ohio who charged $3,200 to his card soon found his debt to be $10,700 because of penalties, fees, and interest.

These were not boiler-room operators charging high interest rates and fees, but some of the biggest and presumably most reputable banks in America—those whose advertising campaigns would make you believe that you and the bank were "family." Would a family member steal your wallet? No. But these banks, with a transaction somewhat removed from cash, apparently would.

Once you view dishonesty through this lens, it is clear that you can't open a newspaper in the morning without seeing new examples to add.

AND SO WE return to our original observation: isn't cash strange? When we deal with money, we are primed to think about our actions as if we had just signed an honor code. If you look at a dollar bill, in fact, it seems to have been designed to conjure up a contract: THE UNITED STATES OF AMERICA, it says in prominent type, with a shadow beneath that makes it seem three-dimensional. And there is George Washington himself (and we all know that he could never tell a lie). And then, on the back, it gets even more serious: IN GOD WE TRUST, it says. And then we've got that weird pyramid, and on top, that unblinking eye! And it's looking right at us! In addition to all this symbolism, the sanctity of money could also be aided by the fact that money is a clear unit of exchange. It's hard to say that a dime is not a dime, or a buck isn't a buck.

But look at the latitude we have with nonmonetary exchanges. There's always a convenient rationale. We can take a pencil from work, a Coke from the fridge—we can even backdate our stock options—and find a story to explain it all. We can be dishonest without thinking of ourselves as dishonest. We can steal while our conscience is apparently fast asleep.

How can we fix this? We could label each item in the supply cabinet with a price, for instance, or use wording that explains stocks and stock options clearly in terms of their monetary value. But in the larger context, we need to wake up to the connection between nonmonetary currency and our tendency to cheat. We need to recognize that once cash is a step away, we will cheat by a factor bigger than we could ever imagine. We need to wake up to this—individually and as a nation, and do it soon.

Why? For one thing, the days of cash are coming to a close. Cash is a drag on the profits of banks—they want to get rid of it. On the other hand, electronic instruments are very profitable. Profits from credit cards in the United States rose from $9 billion in 1996 to a record $27 billion in 2004. By 2010, banking analysts say, there will be $50 billion in new electronic transactions, nearly twice the number processed under the Visa and MasterCard brands in 2004.[26] The question, therefore, is how we can control our tendency to cheat when we are brought to our senses only by the sight of cash—and what we can do now that cash is going away.

Willie Sutton allegedly said that he robbed banks because that's where the money was. By that logic he might be writing the fine print for a credit card company today or penciling in blackout dates for an airline. It might not be where the cash is, but it's certainly where you will find the money.

CHAPTER 13

Beer and Free Lunches

What Is Behavioral Economics, and Where Are the Free Lunches?

The Carolina Brewery is a hip bar on Franklin Street, the main street outside the University of North Carolina at Chapel Hill. A beautiful street with brick buildings and old trees, it has many restaurants, bars, and coffee shops—more than one would expect to find in a small town.

As you open the doors to the Carolina Brewery, you see an old building with high ceilings and exposed beams, and a few large stainless steel beer containers that promise a good time. There are semiprivate tables scattered around. This is a favorite place for students as well as for an older crowd to enjoy good beer and food.

Soon after I joined MIT, Jonathan Levav (a professor at Columbia) and I were mulling over the kinds of questions one might conjure up in such a pleasant pub. First, does the sequential process of taking orders (asking each person in turn to state his or her order) influence the choices that the people sitting around the table ultimately make? In other words, are the patrons influenced by the selections of the others around them? Second, if this is the case, does it encourage conformity or nonconformity? In other words, would the patrons sitting around a table intentionally choose beers that were different from or the same as the choices of those ordering before them? Finally, we wanted to know whether being influenced by others' choices would make people better or worse off, in terms of how much they enjoyed their beer.

THROUGHOUT THIS BOOK, I have described experiments that I hoped would be surprising and illuminating. If they were, it was largely because they refuted the common assumption that we are all fundamentally rational. Time and again I have provided examples that are contrary to Shakespeare's depiction of us in "What a piece of work is a man." In fact, these examples show that we are not noble in reason, not infinite in faculty, and rather weak in apprehension. (Frankly, I think Shakespeare knew that very well, and this speech of Hamlet's is not without irony.)

In this final chapter, I will present an experiment that offers one more example of our predictable irrationality. Then I will further describe the general economic perspective on human behavior, contrast it with behavioral economics, and draw some conclusions. Let me begin with the experiment.

To GET TO the bottom of the sudsy barrel of questions that we thought of at the Carolina Brewery, Jonathan and I decided to plunge in—metaphorically, of course. We started by asking the manager of the Carolina Brewery to let us serve free samples of beer to the customers—as long as we paid for the beer ourselves. (Imagine how difficult it was, later, to convince the MIT accountants that a $1,400 bill for beer is a legitimate research expense.) The manager of the bar was happy to comply. After all, he would sell us the beer and his customers would receive a free sample, which would presumably increase their desire to return to the brewery.

Handing us our aprons, he established his one and only condition: that we approach the people and get their orders for samples within one minute of the time they sat down. If we couldn't make it in time, we would indicate this to the regular waiters and they would approach the table and take the orders. This was reasonable. The manager didn't know how efficient we could be as waiters, and he didn't want to delay the service by too much. We started working.

I approached a group as soon as they sat down. They seemed to be undergraduate couples on a double date. Both guys were wearing what looked like their best slacks, and the girls had on enough makeup to make Elizabeth Taylor look unadorned in comparison. I greeted them, announced that the brewery was offering free beer samples, and then proceeded to describe the four beers:

(1) Copperline Amber Ale: A medium-bodied red ale with a well-balanced hop and malt character and a traditional ale fruitiness.
(2) Franklin Street Lager: A Bohemian pilsner-style golden lager brewed with a soft maltiness and a crisp hoppy finish.
(3) India Pale Ale: A well-hopped robust ale originally brewed to withstand the long ocean journey from England around the Cape of Good Hope to India. It is dry-hopped with cascade hops for a fragrant floral finish.
(4) Summer Wheat Ale: Bavarian-style ale, brewed with 50

percent wheat as a light, spritzy, refreshing summer drink. It is gently hopped and has a unique aroma reminiscent of banana and clove from an authentic German yeast strain.

Which would you choose?

☐ Copperline Amber Ale
☐ Franklin Street Lager
☐ India Pale Ale
☐ Summer Wheat Ale

After describing the beers, I nodded at one of the guys— the blond-haired guy—and asked for his selection; he chose the India Pale Ale. The girl with the more elaborate hairdo was next; she chose the Franklin Street Lager. Then I turned to the other girl. She opted for the Copperline Amber Ale. Her boyfriend, who was last, selected the Summer Wheat Ale. With their orders in hand, I rushed to the bar, where Bob—the tall, handsome bartender, a senior in computer science—stood smiling. Aware that we were in a hurry, he filled my order before any of the others. I then took the tray with the four two-ounce samples back to the double-daters' table and placed their beers in front of them.

Along with their samples, I handed each of them a short survey, printed on the brewery's stationery. In this survey we asked the respondents how much they liked their beer and whether they had regretted choosing that particular brew. After I collected their surveys, I continued to observe the four people from a distance to see whether any of them took a sip of anyone else's beer. As it turned out, none of them shared a sample.

Jonathan and I repeated this procedure with 49 more tables. Then we continued, but for the next 50 tables we
changed the procedure. This time, after we read the descrip-
tions of the beers, we handed the participants a small menu
with the names of the four beers and asked each of them to
write down their preferred beer, rather than simply say it out
loud. In so doing, we transformed ordering from a public
event into a private one. This meant that each participant
would not hear what the others—including, perhaps, someone they were trying hard to impress—ordered and so could
not be influenced by it.

What happened? We found that when people order out
loud in sequence, they choose differently from when they order in private. When ordering sequentially (publicly), they
order more types of beer per table—in essence opting for variety. A basic way to understand this is by thinking about the
Summer Wheat Ale. This brew was not very attractive to
most people. But when the other beers were "taken," our
participants felt that they had to choose something different—
perhaps to show that they had a mind of their own and
weren't trying to copy the others—and so they chose a different beer, one that they may not have initially wanted, but one
that conveyed their individuality.

What about their enjoyment of the beer? It stands to reason that if people choose beer that nobody has chosen just to
convey uniqueness, they will probably end up with a beer
that they don't really want or like. And indeed this was the
case. Overall, those who made their choices out loud, in the
standard way that food is ordered at restaurants, were not as
happy with their selections as those who made their choices
privately, without taking others' opinions into consideration.
There was, however, one very important exception: the first

person to order beer in the group that made its decisions out loud was de facto in the same condition as the people who expressed their opinion privately, since he or she was unencumbered, in choosing, by other people's choices. Accordingly, we found that the first person to order beer in the sequential group was the happiest of his or her group and just as happy as those who chose their beers in private.

BY THE WAY, a funny thing happened when we ran the experiment in the Carolina Brewery: Dressed in my waiter's outfit, I approached one of the tables and began to read the menu to the couple there. Suddenly, I realized that the man was Rich, a graduate student in computer science, someone with whom I had worked on a project related to computational vision three or four years earlier. Because the experiment had to be conducted in the same way each time, this was not a good time for me to chat with him, so I put on a poker face and launched into a matter-of-fact description of the beers. After I finished, I nodded to Rich and asked, "What can I get you?" Instead of giving me his order, he asked how I was doing.

"Very well, thank you," I said. "Which of the beers can I get you?"

He and his companion both selected beers, and then Rich took another stab at conversation: "Dan, did you ever finish your PhD?"

"Yes," I said, "I finished about a year ago. Excuse me; I will be right back with your beers." As I walked to the bar to fill their order, I realized that Rich must have thought that this was my profession and that a degree in social science would only get someone a job as a beer server. When I got back to the table with the samples, Rich and his companion—

who was his wife—tasted the beers and answered the short questionnaire. Then Rich tried again. He told me that he had recently read one of my papers and liked it a lot. It was a good paper, and I liked it, too, but I think he was just trying to make me feel better about my job as a beer server.

ANOTHER STUDY, CONDUCTED later at Duke with wine samples and MBA students, allowed us to measure some of the participants' personality traits—something the manager of the Carolina Brewery had not been thrilled about. That opened the door for us to find out what might be contributing to this interesting phenomenon. What we found was a correlation between the tendency to order alcoholic beverages that were different from what other people at the table had chosen and a personality trait called "need for uniqueness." In essence, individuals more concerned with portraying their own uniqueness were more likely to select an alcoholic beverage not yet ordered at their table in an effort to demonstrate that they were in fact one of a kind.

What these results show is that people are sometimes willing to sacrifice the pleasure they get from a particular consumption experience in order to project a certain image to others. When people order food and drinks, they seem to have two goals: to order what they will enjoy most and to portray themselves in a positive light in the eyes of their friends. The problem is that once they order, say, the food, they may be stuck with a dish they don't like—a situation they often regret. In essence, people, particularly those with a high need for uniqueness, may sacrifice personal utility in order to gain reputational utility.

Although these results were clear, we suspected that in other cultures—where the need for uniqueness is not considered a positive trait—people who ordered aloud in public would try to portray a sense of belonging to the group and express more conformity in their choices. In a study we conducted in Hong Kong, we found that this was indeed the case. In Hong Kong, individuals also selected food that they did not like as much when they selected it in public rather than in private, but these participants were more likely to select the same item as the people ordering before them—again making a regrettable mistake, though a different type of mistake, when ordering food.

FROM WHAT I have told you so far about this experiment, you can see that a bit of simple life advice—a free lunch—comes out of this research. First, when you go to a restaurant, it's a good idea to plan your order before the waiter approaches you, and stick to it. Being swayed by what other people choose might lead you to choose a worse alternative. If you're afraid that you might be swayed anyway, a useful strategy is to announce your order to the table before the waiter comes. This way, you have staked a claim to your order, and it's less likely that the other people around the table will think you are not unique, even if someone else orders the same dish before you get your chance. But of course the best option is to order first.

Perhaps restaurant owners should ask their customers to write out orders privately (or quietly give their orders to the waiters), so that no customer will be influenced by the orders of his or her companions. We pay a lot of money for the pleasure of dining out. Getting people to order anonymously is most likely the cheapest and simplest way to increase the enjoyment derived from these experiences.

But there's a bigger lesson that I would like to draw from this experiment—and in fact from all that I have said in the preceding chapters. Standard economics assumes that we are rational—that we know all the pertinent information about our decisions, that we can calculate the value of the different options we face, and that we are cognitively unhindered in weighing the ramifications of each potential choice.

The result is that we are presumed to be making logical and sensible decisions. And even if we make a wrong decision from time to time, the standard economics perspective suggests that we will quickly learn from our mistakes either on our own or with the help of "market forces." On the basis of these assumptions, economists draw far-reaching conclusions about everything from shopping trends to law to public policy.

But, as the results presented in this book (and others) show, we are all far less rational in our decision making than standard economic theory assumes. Our irrational behaviors are neither random nor senseless—they are systematic and predictable. We all make the same types of mistakes over and over, because of the basic wiring of our brains. So wouldn't it make sense to modify standard economics and move away from naive psychology, which often fails the tests of reason, introspection, and—most important—empirical scrutiny?

Wouldn't economics make a lot more sense if it were based on how people actually behave, instead of how they should behave? As I said in the Introduction, that simple idea is the basis of behavioral economics, an emerging field focused on the (quite intuitive) idea that people do not always behave rationally and that they often make mistakes in their decisions.

In many ways, the standard economic and Shakespearean views are more optimistic about human nature, since they assume that our capacity for reasoning is limitless. By the same token the behavioral economics view, which acknowledges human deficiencies, is more depressing, because it demonstrates the many ways in which we fall short of our ideals. Indeed, it can be rather depressing to realize that we all continually make irrational decisions in our personal, professional, and social lives. But there is a silver lining: the fact that we make mistakes also means that there are ways to improve our decisions—and therefore that there are opportunities for "free lunches."

ONE OF THE main differences between standard and behavioral economics involves this concept of "free lunches." According to the assumptions of standard economics, all human decisions are rational and informed, motivated by an accurate concept of the worth of all goods and services and the amount of happiness (utility) all decisions are likely to produce. Under this set of assumptions, everyone in the marketplace is trying to maximize profit and striving to optimize his experiences. As a consequence, economic theory asserts that there are no free lunches—if there were any, someone would have already found them and extracted all their value.

Behavioral economists, on the other hand, believe that people are susceptible to irrelevant influences from their immediate environment (which we call context effects), irrelevant emotions, shortsightedness, and other forms of irrationality (see any chapter in this book or any research paper in behavioral economics for more examples). What good news can accompany this realization? The good news is that these mistakes also provide opportunities for improvement. If we all make systematic mistakes in our decisions, then why not develop new strategies, tools, and methods to help us make better decisions and improve our overall well-being? That's exactly the meaning of free lunches from the perspective of behavioral economics—the idea that there are tools, methods, and policies that can help all of us make better decisions and as a consequence achieve what we desire.

For example, the question why Americans are not saving enough for retirement is meaningless from the perspective of standard economics. If we are all making good, informed decisions in every aspect of our lives, then we are also saving the exact amount that we want to save. We might not save much because we don't care about the future, because we are looking forward to experiencing poverty at retirement, because we expect our kids to take care of us, or because we are hoping to win the lottery—there are many possible reasons. The main point is that from the standard economic perspective, we are saving exactly the right amount in accordance with our preferences.

But from the perspective of behavioral economics, which does not assume that people are rational, the idea that we are not saving enough is perfectly reasonable. In fact, research in behavioral economics points to many possible reasons why people are not saving enough for retirement. People procrastinate. People have a hard time understanding the real cost of not saving as well as the benefits of saving. (By how much would your life be better in the future if you were to deposit an additional $1,000 in your retirement account every month for the next 20 years?) Being "house rich" helps people believe that they are indeed rich. It is easy to create consumption habits and hard to give them up. And there are many, many more reasons.

The potential for free lunches from the perspective of

behavioral economics lies in new methods, mechanisms, and other interventions that would help people achieve more of what they truly want. For example, the new and innovative credit card that I described in Chapter 6, on self-control, could help people exercise more self-control within the domain of spending. Another example of this approach is a mechanism called "save more tomorrow," which Dick Thaler and Shlomo Benartzi proposed and tested a few years ago.

Here's how "save more tomorrow" works. When new employees join a company, in addition to the regular decisions they are asked to make about what percentage of their paycheck to invest in their company's retirement plan, they are also asked what percentage of their future salary raises they would be willing to invest in the retirement plan. It is difficult to sacrifice consumption today for saving in the distant future, but it is psychologically easier to sacrifice consumption in the future, and even easier to give up a percentage of a salary increase that one does not yet have.

When the plan was implemented in Thaler and Benartzi's test, the employees joined and agreed to have their contribution, as a percentage, increase with their future salary raises. What was the outcome? Over the next few years, as the employees received raises, the saving rates increased from about 3.5 percent to around 13.5 percent—a gain for the employees, their families, and the company, which by now had more satisfied and less worried employees.

This is the basic idea of free lunches—providing benefits for all the parties involved. Note that these free lunches don't have to be without cost (implementing the self-control credit card or "save more tomorrow" inevitably involves a cost). As long as these mechanisms provide more benefits than costs, we should consider them to be free lunches—mechanisms that provide net benefits to all parties.

IF I WERE to distill one main lesson from the research described in this book, it is that we are pawns in a game whose forces we largely fail to comprehend. We usually think of ourselves as sitting in the driver's seat, with ultimate control over the decisions we make and the direction our life takes; but, alas, this perception has more to do with our desires—with how we want to view ourselves—than with reality.

Each of the chapters in this book describes a force (emotions, relativity, social norms, etc.) that influences our behavior. And while these influences exert a lot of power over our behavior, our natural tendency is to vastly underestimate or completely ignore this power. These influences have an effect on us not because we lack knowledge, lack practice, or are weak-minded. On the contrary, they repeatedly affect experts as well as novices in systematic and predictable ways. The resulting mistakes are simply how we go about our lives, how we "do business." They are a part of us.

Visual illusions are also illustrative here. Just as we can't help being fooled by visual illusions, we fall for the "decision illusions" our minds show us. The point is that our visual and decision environments are filtered to us courtesy of our eyes, our ears, our senses of smell and touch, and the master of it all, our brain. By the time we comprehend and digest information, it is not necessarily a true reflection of reality. Instead, it is our representation of reality, and this is the input we base our decisions on. In essence we are limited to the tools nature has given us, and the natural way in which we make decisions is limited by the quality and accuracy of these tools.

A second main lesson is that although irrationality is commonplace, it does not necessarily mean that we are helpless. Once we understand when and where we may make erroneous decisions, we can try to be more vigilant, force ourselves to think differently about these decisions, or use technology to overcome our inherent shortcomings. This is also where businesses and policy makers could revise their thinking and consider how to design their policies and products so as to provide free lunches.

———

THANK YOU FOR reading this book. I hope you have gained some interesting insights about human behavior, gained some insight into what really makes us tick, and discovered ways to improve your decision making. I also hope that I have been able to share with you my enthusiasm for the study of rationality and irrationality. In my opinion, studying human behavior is a fantastic gift because it helps us better understand ourselves and the daily mysteries we encounter. Although the topic is important and fascinating, it is not easy to study, and there is still a lot of work ahead of us. As the Nobel laureate Murray Gell-Mann once said, "Think how hard physics would be if particles could think."

Irrationally yours,
Dan Ariely

PS: If you want to participate in this journey, log on to www.predictablyirrational.com, sign up for a few of our studies, and leave us your ideas and thoughts.

Reflections and Anecdotes about Some of the Chapters

Reflections on Dating and Relativity (Chapter 1)

In Chapter 1, on relativity, I offered some dating advice. I proposed that if you want to go bar-hopping, you should consider taking along someone who looks similar to you but who is slightly less attractive than you are. Because of the relative nature of evaluations, others would perceive you not only as cuter than your decoy, but also as better-looking than other people in the bar. By the same logic, I also pointed out that the flip side of this coin is that if someone invites you to be his or her wingman (or wingwoman), you can easily figure out what your friend really thinks of you. As it turns out, I forgot to include one important warning that came courtesy of the daughter of a colleague of mine from MIT.

"Susan" was an undergraduate at Cornell who wrote to me, saying she was delighted with my trick and that it had worked wonderfully for her. Once she found the ideal decoy, her social life improved. But a few weeks later she wrote again, telling me that she'd been at a party where she'd had a few drinks. For some odd reason, she decided to tell her friend why she invited her to accompany her everywhere. The friend was understandably upset, and the story did not end well.

The moral of this story? Never, ever tell your friend why you're asking him or her to come with you. Your friend might have suspicions, but for the love of God, don't eliminate all doubt.

Reflections on Traveling and Relativity (Chapter 1)

When *Predictably Irrational* came out, I went on a book tour that lasted six straight weeks. I traveled from airport to airport, city to city, radio station to radio station, talking to reporters and readers for what seemed like days on end, without engaging in any type of personal discussion. Every conversation was short, "all business," and focused on my research. There was no time to enjoy a cup of coffee or a beer with any of the wonderful people I encountered.

Toward the end of the tour I found myself in Barcelona. There I met Jon, an American tourist who, like me, did not speak any Spanish. We felt an immediate camaraderie. I imagine this kind of bonding happens often with travelers from the same country who are far from home and find themselves sharing observations about how they differ from the locals around them. Jon and I ended up having a wonderful dinner and a deeply personal discussion. He told me things that he seemed not to have shared before, and I did the same. There was an unusual closeness between us, as if we were long-lost brothers. After staying up very late talking, we both needed to sleep. We would not have a chance to meet again before parting ways the following morning, so we exchanged e-mail addresses. This was a mistake.

About six months later, Jon and I met again for lunch in New York. This time, it was hard for me to figure out why I'd felt such a connection with him, and no doubt he felt the same. We had a perfectly amicable and interesting lunch, but it lacked the intensity of our first meeting, and I was left wondering why.

In retrospect, I think it was because I'd fallen victim to the effects of relativity. When Jon and I first met, everyone around us was Spanish, and as cultural outsiders we were each other's best alternative for companionship. But once we returned home to our beloved American families and friends, the basis for comparison switched back to "normal" mode. Given this situation, it was hard to understand why Jon or I would want to spend another evening in each other's company rather than with those we love.

My advice? Understand that relativity is everywhere, and that we view everything through its lens—rose-colored or otherwise. When you meet someone in a different country or city and it seems that you have a magical connection, realize that the enchantment might be limited to the surrounding circumstances. This realization might prevent you from subsequent disenchantment.

Reflections on the Price of FREE! (Chapter 3)

We learned from our experiments that we all get a bit too excited when something is FREE! and that consequently, we can make decisions that are not in our best interest.

For example, imagine that you were choosing between two credit cards: one that offers you a 12 percent APR but has no yearly fee (FREE!), and one that offers you a lower interest rate of 9 percent APR but charges you a $100 annual fee. Which one would you take? Most people would overemphasize the yearly fee and in pursuit of the FREE! offer would end up getting the card that costs them much more in the long run—when they inevitably miss a payment or carry a balance.*

Although identifying and fighting the allure of FREE! is important in order to avoid traps while we are making decisions, there are also some cases in which we can use FREE! to our advantage. Take, for example, the common experience of going to a restaurant with friends. When the server drops off the check at the end of a meal, people often scramble to figure out the norms for payment. Do we each pay for what we ordered? Do we split the bill evenly, even if John had that extra glass of wine and the crème brûlée? FREE! can help us solve this problem, and in the process help us get more joy from dining out with our friends.

The answer, as it turns out, is that one person should pay the entire bill, and that the people involved should take turns paying over time. Here is the logic: When we pay—regardless of the amount of money—we feel some psychological pain, which social scientists call the "pain of paying." This is the unpleasantness associated with giving up our hard-earned cash, regardless of the circumstances. It turns out that the pain of paying has two interesting features. First, and most obviously, when we pay nothing (for example, when someone else foots the bill) we don't feel any pain of paying. Second, and less obviously, the pain of paying is relatively insensitive

---

*When it comes to credit cards, the appeal of FREE! is further enhanced because most of us are overoptimistic about our financial future, and overconfident about our ability to always pay our bills on time.

to the amount that we pay. This means that we feel more pain of paying as the bill increases, but every additional dollar on the bill pains us less. (We call this "diminishing sensitivity." Analogously, if you add one pound to an empty backpack, it feels like a substantial increase in weight. But adding a pound to a backpack that's already laden with a laptop and some books does not feel like a big difference.) This diminishing sensitivity to the pain of paying means that the first dollar we pay will cause us the highest pain, the second dollar will cause us less, and so on, until we feel just a tiny twinge for, say, the forty-seventh dollar.

So if we are dining with others, we are happiest when we pay nothing (FREE!); we are less happy when we have to pay something; and the additional dollars we fork over cause us a smaller and smaller additional amount of pain as the size of the bill increases. The logical conclusion is that one person should pay the whole bill.

If you're still unconvinced, consider the following example: Imagine that four people share a meal and the bill comes to $100. Now, if everyone at the table pays $25, every person would feel some pain of paying. In order to make this less abstract, let's assign "units" as a measure of this pain. We'll assume that paying $25 translates into 10 units of pain for a total of 40 units of pain for the whole table when it comes time to split the bill. But what if one person pays the entire bill? Since the pain of paying does not increase linearly with the amount of payment, the person who is paying will feel 10 units of pain for the first $25 that he or she pays; maybe 7 units for the next $25; 5 units for the next $25; and 4 units for the last $25. The total of 26 units of pain lowers the amount of pain for the entire table by 14 units. The general point is this: we all love getting our meals for nothing, and as long as we can alternate payers, we can enjoy many FREE! dinners and derive greater overall benefit from our friendships in the process.

"Aha," you might say, "but what about times when I eat only a green salad while my friend's husband orders a green salad, a filet mignon dinner, two glasses of the most expensive cabernet sauvignon, and a crème brûlée for dessert? Or when the number of people changes the next time we gather? Or when some people in the group leave town altogether? All of this leaves me holding the bag."

Certainly, there is no question that all these considerations make the "I'll buy this time, you buy next time" approach less economically efficient. Nevertheless, given the huge benefits in terms of the pain of paying that this method delivers, I personally would be willing to sacrifice a few bucks here and there to reduce the pain of paying for my friends and myself.

Reflections on Social Norms: Lessons on Gifts (Chapter 4)

When we mix social and monetary norms, strange and undesirable things can happen. For example, if you walk your date home after a wonderful evening together, don't mention how much the evening set you back. That is not a good strategy for getting a passionate good-night kiss. (I certainly do not recommend this as an experiment, but if you do try it, let me know how it turns out.) Dating, of course, is just one arena in which we can mess up a social relationship by introducing financial norms, and this danger lurks around many corners.

On some level we all know this, and therefore we sometimes deliberately make decisions that do not fall into line with rational economic theory. Think of gifts, for example. From a standard economic perspective, they are a waste of money. Imagine that you invite me to your home for dinner one evening, and I decide to spend $50 on a nice bottle of Bordeaux as a token of gratitude. There are some problems with this decision: You might not like Bordeaux. You might have preferred something else: a copy of *Predictably Irrational*, a DVD of *Citizen Kane*, or a blender. This means that the bottle of wine that cost me $50 might be worth, at most, $25 to you in terms of its utility. That is, for $25 you could get something else that would make you just as happy as my $50 bottle of wine.

Now, if giving gifts was a rational activity, I would come to dinner and say, "Thank you for inviting me for dinner. I was going to spend $50 on a bottle of Bordeaux, but I realize that this might provide you with far less happiness than $50 in cash." I peel off five $10 bills, hand them to you, and add, "Here you go. You can decide how best to spend it." Or maybe I would give you $40 in cash and make us both better off—not to mention saving myself the trouble of shopping for wine.

Although we all realize that offering cash instead of gifts is more economically efficient, I don't expect that many people will follow this rational advice, because we all know that doing so will in no way endear us to our hosts. If you want to demonstrate affection, or strengthen your relationship, then giving a gift—even at the risk that it won't be appreciated as much as you hoped—is the only way to go.

So imagine two scenarios. Let's say it's the holidays, and two different neighbors invite you to their parties in the same week. You accept both invitations. In one case, you do the irrational thing and give Neighbor X a bottle of Bordeaux;

for the second party you adopt the rational approach and give Neighbor Z $50 in cash. The following week, you need some help moving a sofa. How comfortable would you be approaching each of your neighbors, and how do you think each would react to your request for a favor? The odds are that Neighbor X will step in to help. And Neighbor Z? Since you have already paid him once (to make and share dinner with you), his logical response to your request for help might be, "Fine. How much will you pay me this time?" Again, the prospect of acting rationally, financially speaking, sounds deeply irrational in terms of social norms.

The point is that while gifts are financially inefficient, they are an important social lubricant. They help us make friends and create long-term relationships that can sustain us through the ups and downs of life. Sometimes, it turns out, a waste of money can be worth a lot.

Reflections on Social Norms: Benefits in the Workplace (Chapter 4)

The same general principles regarding social norms also apply to the workplace. In general, people work for a paycheck, but there are other, intangible benefits we get from our jobs. These are also very real and very important, yet much less understood.

Often, when I'm on a flight and the people sharing the row with me don't immediately put their headphones on, I enter into an interesting discussion with one of them. Invariably, I learn a lot about my neighbor's work, work history, and future projects. On the other hand, I discover very little about the person's family, favorite music, movies, or hobbies. Unless my neighbor gives me a business card, I almost never learn his or her name until the moment we are both about to leave the plane. There are probably many reasons for this, but I suspect that one of them is that most people take a lot of pride in their work. Of course this may not be true of everyone, but I think that for many people the workplace is not just a source of money but also a source of motivation and self-definition.

Such feelings benefit both the workplace and the employee. Employers who can foster these feelings gain dedicated, motivated employees who think about solving job-related problems even after the workday is over. And employees who take pride in their work feel a sense of happiness and purpose. But in the same way that market norms may undermine social norms, it may be that market norms also erode the pride and meaning people get from the workplace (for example, when we pay schoolteachers according to their students' performance on standardized tests).

Imagine that you work for me, and that I want to give you a year-end bonus. I offer you a choice: $1,000 in cash or an all-expenses-paid weekend in the Bahamas, which would cost me $1,000. Which option would you choose? If you are like most people who have answered this question, you would take the cash. After all, you may have already been to the Bahamas and may not have enjoyed being there very much, or maybe you'd prefer to spend a weekend at a resort closer to home and use the remainder of the bonus money to buy a new iPod. In either case, you think that you can best decide for yourself how to spend the money.

This arrangement seems to be financially efficient, but would it increase your happiness with your work, or your loyalty to the company? Would it make me a better boss? Would it improve the employer-employee relationship in any way? I suspect that both your and my best interests would be better served if I simply didn't offer you a choice and just sent you on the Bahamas vacation. Consider how much more relaxed and refreshed you would feel, and how well you would perform, after a relaxing weekend of sun and sand, compared with how you would feel and behave after you got the $1,000 bonus. Which would help you feel more committed to your job, more enjoyment in your work, more dedication to your boss? Which gift would make you more likely to stay long hours one night to meet an important deadline? On all of these, the vacation beats the cash hands down.

This principle doesn't apply only to gifts. Many employers, in an effort to show their employees how well they are treating them, add different line items to their paycheck stubs, describing exactly how much money the employer is spending on health care, retirement plans, the gym at work, and the cafeteria. These items are all legitimate, and they reflect real costs to the employer, but overtly stating the prices of these items changes the workplace from a social environment in which the employer and employee have a deep commitment to each other to a transactional relationship. Explicitly stating the financial value of these benefits can also diminish enjoyment, motivation, and loyalty to the workplace—negatively affecting both the employer-employee relationship and our own pride and happiness at work.

Gifts and employee benefits seem, at first glance, to be an odd and inefficient way of allocating resources. But with the understanding that they fulfill an important role in creating long-term relationships, reciprocity, and positive feelings, companies should try to keep benefits and gifts in the social realm.

Reflections on Immediate Gratification and Self-Control (Chapters 5 and 6)

Oscar Wilde once said, "I never put off till tomorrow what I can do the day after." He seemed to accept and even embrace the role of procrastination in his life, but most of us find the allure of immediate gratification so strong that it wrecks our best-laid plans for dieting, saving money, cleaning the house—the list is endless.

When we have problems with self-control, sometimes we delay tasks that we should do immediately. But we also exhibit problems with self-control when we attend too frequently to tasks that we should put off—such as obsessively checking our e-mail.

The danger of continually checking e-mail was crucial in the plotline of the movie *Seven Pounds*: Will Smith's character checks his phone for e-mail while driving and veers head-on into an oncoming van, killing his wife and six other people. This is just a movie, of course, but compulsively checking e-mail while driving is more common than most of us would care to admit (go ahead, raise your hand*).

I hope that you're not that addicted to e-mail, but too many of us suffer from an unhealthy attachment to it. A recent Australian report found that workers spent an average of 14.5 hours, or more than two working days a week, checking, reading, arranging, deleting, and responding to e-mail.[27] Add to this the rise of social networks and news groups, and you can most likely double the time we spend in virtual interaction and message management.

---

*I teach about 200 graduate students each year, and in early 2009 I asked for a show of hands to the question of how many students had ever used e-mail or text messaging while driving. All but three raised their hands (and one of the three who did not was visually impaired!).

I, for one, have very mixed feelings about e-mail. On one hand, it lets me communicate with colleagues and friends all over the world without the delays of snail mail or the constraints of talking on the phone. (Is it too late to call? What time is it in Auckland anyway?) On the other hand, I receive hundreds of messages a day, including many involving things I don't really care about (announcements, minutes of meetings, and so on). Regardless of whether I care, the ongoing stream of e-mail is a constant distraction.

I once tried to overcome this distraction by resolving to check e-mail only at night, but I quickly discovered that this would not do. Other people expected me to do as they do—check e-mail constantly and rely on it as a sole means of communication. As a result of not checking my e-mail regularly, I ended up going to meetings that had been canceled, or arriving at the wrong time or place. So I gave in, and now I check e-mail way too often, and as I do I constantly sort the messages into categories: spam and unimportant e-mail that I delete right away; messages I might care about or need to respond to at some point in the future; messages I need to respond to immediately; and so on.

In bygone days the mail cart came around the office once or twice a day with a few letters and memos—not so with e-mail, which never takes a break. For me, the day goes like this: I start working on something and get deeply into it. Eventually I get stuck on some difficult point, and decide to take a quick break—obviously, to check e-mail. Twenty minutes later I get back to the task, with little recollection of where I was and what I was thinking. By the time I'm back on track, I've lost both time and some of my focus, and this outcome assuredly does not help me solve whatever problem caused me to take five in the first place.

Sadly, this is not where the story ends. Enter smart phones—an even greater time sink. A while ago I got one of these lovely, distracting gadgets in the form of an iPhone, which meant that I could also check e-mail while waiting in a checkout line, walking into the office, riding in the elevator, while listening to other people's lectures (I haven't yet figured out how to do this during my own lectures), and even while sitting at traffic lights. In truth, the iPhone has made the level of my addiction very clear. I check it almost ceaselessly. (Businesspeople recognize the addictive properties of these devices: this is why they often call their BlackBerries "Crack-Berries.")

I THINK E-MAIL addiction has something to do with what the behavioral psychologist B. F. Skinner called "schedules of re-inforcement." Skinner used this phrase to describe the rela-tionship between actions (in his case, a hungry rat pressing a lever in a so-called Skinner box) and their associated rewards (pellets of food). In particular, Skinner distinguished be-tween fixed-ratio schedules of reinforcement and variable-ratio schedules of reinforcement. Under a fixed schedule, a rat received a reward of food after it pressed the lever a fixed number of times—say 100 times. (To make a human com-parison, a used-car dealer might get a $1,000 bonus for every 10 cars sold.) Under the variable schedule, the rat earned the food pellet after it pressed the lever a random number of times. Sometimes it would receive the food after pressing 10 times, and sometimes after pressing 200 times. (Analogously, our used-car dealer would earn a $1,000 bonus after selling an unknown number of cars.)

Thus, under the variable schedule of reinforcement, the

arrival of the reward is unpredictable. On the face of it, one might expect that the fixed schedules of reinforcement would be more motivating and rewarding because the rat (or the used-car dealer) can learn to predict the outcome of his work. Instead, Skinner found that the variable schedules were actually more motivating. The most telling result was that when the rewards ceased, the rats who were under the fixed schedules stopped working almost immediately, but those under the variable schedules kept working for a very long time.

This variable schedule of reinforcement also works wonders for motivating people. It is the magic (or, more accurately, dark magic) that underlies gambling and playing the lottery. How much fun would it be to play a slot machine if you knew in advance that you would always lose nine times before winning once, and that this sequence would continue for as long as you played? It would probably be no fun at all! In fact, the joy of gambling comes from the inability to predict when rewards are coming, so we keep playing.

So, what do food pellets and slot machines have to do with e-mail? If you think about it, e-mail is very much like gambling. Most of it is junk and the equivalent to pulling the lever of a slot machine and losing, but every so often we receive a message that we really want. Maybe it contains good news about a job, a bit of gossip, a note from someone we haven't heard from in a long time, or some important piece of information. We are so happy to receive the unexpected e-mail (pellet) that we become addicted to checking, hoping for more such surprises. We just keep pressing that lever, over and over again, until we get our reward.

This explanation gives me a better understanding of my e-mail addiction, and more important, it might suggest a few means of escape from this Skinner box and its variable schedule of reinforcement. One helpful approach I've discovered is to turn off the automatic e-mail-checking feature. This action doesn't eliminate my checking, but it reduces the frequency with which my computer notifies me that I have new e-mail waiting (some of it, I would think to myself, must be interesting or relevant). Additionally, many applications allow users to link different colors and sounds to different incoming e-mail. For example, I assign every e-mail on which I'm cc'd to the color gray, and send it directly to a folder labeled "Later." Similarly, I set my application to play a particularly cheerful sound when I receive a message from a source I've marked as urgent and important (these include messages from my wife, students, or members of my department). Sure, it takes some time to set up such filters, but having once gone to the trouble of doing so, I've reduced the randomness of the reward, made the schedule of reinforcement more fixed, and ultimately improved my life. As for overcoming the temptations of checking my iPhone too frequently—I am still working on that one.

Further Reflection on Self-Control:
The Lesson of Interferon (Chapters 5 and 6)

Several years ago I heard an interview on NPR with the Delany sisters, who lived to be 102 and 104. There was one particular part of the interview that remained with me. The sisters said that one of their secrets for a long life was never marrying, because they never had husbands to "worry them to death." This sounds reasonable enough, but it isn't something to which I can personally attest (and it also turns out that men benefit more from marriage anyway).[28] One of the sisters said that another secret was to avoid hospitals, which seemed sensible for two reasons—if you're healthy in the first place, you don't need to go, and you're also less likely to catch an illness from being in the hospital.

I certainly understood what she was talking about. When I was first hospitalized for my burns, I acquired hepatitis from a blood transfusion. Obviously, there's no good time to get hepatitis, but the timing could not have been worse for me. The disease increased the risks of my operations, delayed my treatment, and caused my body to reject many of the skin transplants. After a while the hepatitis subsided, but it still slowed my recovery by flaring up from time to time and wreaking havoc on my system.

This was in 1985, before my type of hepatitis had been isolated; the doctors knew it wasn't hepatitis A or B, but it remained a mystery, so they just called it non-A-non-B hepatitis. In 1993, when I was in graduate school, I had a flare-up; I checked into the student health center and the doctor told me I had hepatitis C, which had recently been isolated and identified. This was good news for two reasons. First, I now knew what I had, and second, a new experimental treatment, interferon, looked promising. Given the threat of liver fibrosis, cirrhosis, and the possibility of early death from hepatitis C, it seemed to me that being part of an experimental study was clearly the lesser of two evils.

Interferon was initially approved by the FDA for treatment of hairy cell leukemia (which has no other real treatment) and, as is often the case with cancer therapy, the treatment regimen was particularly distasteful. The initial protocol called for self-injections of interferon three times a week. I was warned that after each injection I would experience fever, nausea, headaches, and vomiting, and this warning was accurate. So, for six months on Mondays, Wednesdays, and Fridays I would arrive home from school, take the needle from the medicine cabinet, open the refrigerator, load the syringe with the right dosage of interferon, and plunge the needle into my thigh. Then I would lie down in the big hammock—the only interesting piece of furniture in my loft-like student apartment—from which I had a perfect view of the television. I kept a bucket within reach to catch the vomit that would inevitably come up, after which the fever, shivering, and headache would begin. At some point I would fall asleep and wake up aching with flulike symptoms. By noon I would be more or less OK and would go back to work.

The difficulty that I, and the rest of the patients, had with the interferon was the basic problem of delayed gratification and self-control. On every injection day I was faced with a trade-off between giving myself an injection and feeling sick for the next 16 hours (a negative immediate effect), and the hope that the treatment would cure me in the long term (a positive long-term effect). At the end of the six-month trial the doctors told me that I was the only patient in the protocol who had followed the regimen in the way they designed it. Everyone else in the study skipped the medication numerous times, which was hardly surprising, given the challenges. (Lack of medical compliance is, in fact, a very pervasive problem.)

So how did I do it? Did I simply have nerves of steel? No. Like anyone else, I have plenty of problems with self-control. But I did have a trick. I basically tried to harness my other desires in an effort to make the prospect of the terrible injection more bearable. For me, the key was movies.

I love movies. If I had the time, I would watch one every day. When the doctors told me what to expect, I decided that

I would not watch any movies until after I injected myself, and then I could watch as many as I wanted until I fell asleep.

On every injection day, I would stop at the video store on the way to school and pick up a few films that I wanted to see. I would have these in my bag and would eagerly antici-pate watching them later that day. Then, immediately after I took the injection, but before the shivering and headache set in, I jumped into my hammock, got comfortable, made sure the bucket was in position, and started my mini–film fest. This way, I learned to associate the initial injection with the rewarding experience of watching a wonderful movie. Only an hour later, after the negative side effects kicked in, did I have a less than wonderful feeling about it.

Planning my evenings in this way helped my brain associ-ate the injection more closely with the movie than with the fever, chills, and vomiting, and thus, I was able to continue the treatment.

DURING THE SIX-MONTH treatment, it looked as though the interferon was working, and my liver function improved dra-matically. Unfortunately, a few weeks after the trial was over, the hepatitis returned, so I started a more aggressive treat-ment. This one lasted a year and involved not only interferon but also a drug called ribavirin. To compel myself to follow this treatment, I again tried the injection-movie-hammock procedure as before. (Thanks to my somewhat faulty mem-ory, I was even able to enjoy some of the same movies I had watched during the first treatment with interferon.)

This time, however, I was also interviewing at various universities for a job as an assistant professor. I had to travel to 14 cities, stay overnight in hotels, give a talk to groups of academics, and then submit to one-on-one interviews with professors and deans. To avoid telling my prospective colleagues about my adventures with interferon and ribavirin, I would insist on a rather strange schedule of interviews. I routinely had to make some excuse about why I arrived early the day before the interview but could not go out for dinner that evening with my hosts. Instead, I would check into the hotel, take out the injection from a little icebox that I carried with me, inject myself, and watch a few movies on the hotel television. The following day I would also try to delay the interviews for a few hours, but once I felt better I would go through the interview as best I could. (Sometimes my procedure worked; sometimes I had to meet people while I still felt wretched.) Fortunately, after I finished my interviews I received excellent news. Not only had I been offered a job, but the combination treatment had eliminated the hepatitis from my liver. I've been hepatitis-free ever since.

THE LESSON I took away from my interferon treatment is a general one: if a particular desired behavior results in an immediate negative outcome (punishment), this behavior will be very difficult to promote, even if the ultimate outcome (in my case, improved health) is highly desirable. After all, that's what the problem of delayed gratification is all about. Certainly, we know that exercising regularly and eating more vegetables will help us be healthier, even if we don't live to be as old as the Delany sisters; but because it is very hard to hold a vivid image of our future health in our mind's eye, we can't keep from reaching for the doughnuts.

In order to overcome many types of human fallibility, I be-

lieve it's useful to look for tricks that match immediate, powerful, and positive reinforcements with the not-so-pleasant steps
we have to take toward our long-term objectives. For me, beginning a movie—before I felt any side effects—helped me to
sustain the unpleasantness of the treatment. As a matter of
fact, I timed everything perfectly. The moment I finished in-
jecting myself, I pressed the Play button. I suspect that had I hit
Play after the side effects kicked in, I would not have been as
successful in winning the tug-of-war. And who knows? Maybe
if I had waited for the side effects to kick in before I started the
movies, I would have created a negative association and would
now enjoy movies less as a consequence.*

ONE OF MY colleagues at Duke University, Ralph Keeney,
recently noted that America's top killer isn't cancer or heart
disease, nor is it smoking or obesity. It's our inability to make
smart choices and overcome our own self-destructive behaviors.[29] Ralph estimates that about half of us will make a life-
style decision that will ultimately lead us to an early grave.
And as if this were not bad enough, it seems that the rate at
which we make these deadly decisions is increasing at an
alarming pace.

I suspect that over the next few decades, real improvements in life expectancy and quality are less likely to be
driven by medical technology than by improved decision
making. Since focusing on long-term benefits is not our natural tendency, we need to more carefully examine the cases in
which we repeatedly fail, and try to come up with some rem-

---

*I did experience such a negative association with eggs. Soon after I was injured, the
doctors fed me thirty raw eggs daily through a feeding tube. To this day, even the smell
of eggs turns me off.

edies for these situations. (For an overweight movie lover, the key might be to enjoy watching a film while walking on the treadmill.) The trick is to find the right behavioral antidote for each problem. By pairing something that we love with something that we dislike but that is good for us, we might be able to harness desire with outcome—and thus overcome some of the problems with self-control we face every day.

Reflections on the Challenges of Ownership (Chapter 7)

In 2007 and 2008, home values across America plummeted as fast as George W. Bush's approval ratings. Each month brought with it more bad news: more foreclosures, more new homes for sale in a stagnant real estate market, and more stories of people who couldn't get mortgages. Results from a study by Zillow.com (a Web site that facilitates home searches and price estimations) illustrated just how strongly this news affected home owners: in the second quarter of 2008, nine out of ten home owners (92 percent) said there had been foreclosures in their local real estate market, and they were concerned that these foreclosures had lowered home values in their neighborhoods. Moreover, four in five home owners (82 percent) did not see much hope for improvement in the real estate market in the near future.

On the face of it, Zillow's research suggested that home-owners had been paying attention to the media, had an idea of what was happening in the economy, and understood that the housing crunch was a reality. But this study also found that these seemingly well-informed people believed that the values of their own homes had not decreased as much. Two out of three home owners (62 percent) believed that the value of their own home had increased or stayed the same, and about half (56 percent) planned to invest in home improvements, even as they watched the housing market collapse around them. What explained the wide gap between their inflated perception of their homes' values and the gloomy market reality?

As we discussed in Chapter 7, ownership fundamentally changes our perspective. In the same way that we think our own kids are more wonderful and special than our friends' and neighbors' children (regardless of whether our children deserve such esteem), we overvalue everything that we own, whether it's a pair of basketball tickets or our domiciles.

But home ownership is even more interesting and complex than, say, the regular case of owning a coffee mug or a pair of baseball tickets—because we invest so much in our houses. Think, for example, about all the changes and tinkering we do to our homes once we move into them. We replace laminate countertops with granite. We take out a wall and install a new window that lets the light shine just so on the dining room table. We paint the living room walls a deep earthy clay color. We change the bathroom tile. We add a porch and install a koi pond in the backyard. Little by little, we make changes here and there until the house feels perfectly tailored to our unique individual tastes, until it expresses our elegant or eclectic sense of style to everyone else. When the neighbors come over, they admire our countertops and light fixtures. But in the end, do other people value the changes we have so lovingly made as much as we do? Do they value these changes at all?

Consider a home owner who compares her own beautifully remodeled house with a similar one down the street that has been languishing on the market for months, or with another that has recently sold for much less than the asking price. In so doing, she understands why the owners of these other homes had such a hard time selling them. They had the laminate and not the granite countertops, no earthy clay paint or light that fell just so on the dining room table. "No wonder those houses didn't sell," she thinks to herself, "they simply are not as nice as mine."

My wife, Sumi, and I also fell victim to this bias. When we worked at MIT, we bought a new house in Cambridge, Massachusetts (the house was originally built in 1890, but it felt new to us). We promptly went about fixing it up. We took down some walls to give the house an open feel, which we loved. We renovated the bathrooms and set up a sauna in the basement. We also converted the carriage house in the garden into a small combination office-apartment. Sometimes we would pack our laundry basket with some wine, food, and clothing, and escape to the carriage house for a "weekend away."

Then, in 2007, we took jobs at Duke University and moved to Durham, North Carolina. We assumed that the housing market would continue to decline, and that it would be in our best interest to sell the Cambridge house as quickly as possible. We also wanted to avoid having to pay for heating, taxes, and a mortgage on two homes.

Many people came to see our beautifully remodeled Cambridge home. They all seemed to appreciate the structure and the feel of the place, but no one put in an offer. People told us that the house was beautiful, but somehow they could not fully appreciate the benefit of the open floor plan. Instead, they wanted something with more privacy. We heard what they said, but it didn't fully register. "Clearly," we said to each other after each set of prospective buyers had come and gone, "those people are just dull and unimaginative, and have no taste. Surely our beautiful, open, airy home will be just right for the perfect someone."

Time passed. We paid double mortgages, double heating bills, and double taxes while the housing market continued to slow down. Many more people came to see the house and left without extending an offer. Eventually Jean, our real estate agent, delivered the bad news to us the way a doctor tells a patient there's something funny looking on his X-ray. "I think," she said slowly, "that if you want to sell the house, you will have to rebuild some walls and reverse some of the changes you have made." Until she said those words, we had not accepted this truth. Despite our disbelief, and still fully convinced of our superior taste, we took the plunge and paid a contractor to re-erect some walls. A few weeks later the house was sold.

In the end, the buyers didn't want our home. They wanted theirs. This was a very expensive lesson, and I certainly wish we had had a better sense of the effect of our modifications on potential buyers.

OUR PROPENSITY TO overvalue what we own is a basic human bias, and it reflects a more general tendency to fall in love with, and be overly optimistic about, anything that has to do with ourselves. Think about it—don't you feel that you are a better-than-average driver, are more likely to be able to afford retirement, and are less likely to suffer from high cholesterol, get a divorce, or get a parking ticket if you overstay your meter by a few minutes? This positivity bias, as psychologists call it, has another name: "The Lake Wobegone

Effect," named after the fictional town in Garrison Keillor's popular radio series *A Prairie Home Companion*. In Lake Wobegone, according to Keillor, "all the women are strong, all the men are good-looking, and all the children are above average."

I don't think we can become more accurate and objective in the way we think about our children and houses, but maybe we can realize that we have such biases and listen more carefully to the advice and feedback we get from others.

Reflections on Expectations: Music and Food (Chapter 9)

Imagine walking into a truck stop off a deserted stretch of Interstate 95 at nine o'clock in the evening. You've been driving for six hours. You are tired and still have a long drive ahead of you. You need a bite to eat and want to be out of the car for a bit, so you walk into what appears to be a restaurant of sorts. It has the usual cracked-vinyl-covered booths and fluorescent lighting. The coffee-stained tabletops leave you a bit wary. Still, you think, "Fine, no one can screw up a hamburger that badly." You reach for the menu, conveniently stashed behind an empty napkin dispenser, only to discover this is no ordinary greasy spoon. Instead of hamburgers and chicken sandwiches, you're astonished to see that the menu offers foie gras au torchon, truffle pâté with frisée and fennel marmalade, gougères with duck confit, quail à la crapaudine, and so on.

Items like this would be no surprise in even a small Manhattan restaurant, of course. And it is possible that the chef got tired of Manhattan, moved to the middle of nowhere, and now cooks for whoever happens through. So is there a key difference between ordering gougères with duck confit in Manhattan and ordering it at an isolated truck stop on I-95? If you encountered such French delicacies at the truck stop, would you be brave enough to try them? Suppose the prices were not listed on the menu. What would you be willing to pay for an appetizer or an entrée? And if you ate it, would you enjoy it as much as you might if you were eating the same food in Manhattan?

On the basis of what we learned from Chapter 9, the answers are simple. Ambience and expectations do add a great deal to our enjoyment. You would expect less in such an environment, and as a consequence you would enjoy the experience at the truck stop less, even if you had the identical foie gras au torchon in both places. Likewise, if you knew that pâté is largely made of run-of-the-mill goose liver and butter* rather than super special ingredients, you would enjoy it much less.

A FEW YEARS ago the folks at the *Washington Post* were curious about the same basic topic and decided to run an experiment.[30] Instead of food, they used music. The experimental question was this: can outstanding art shine through a filter of mundane and dingy expectation?

Journalist Gene Weingarten asked Joshua Bell, generally considered one of the best violinists in the world, to pose as a street performer and play some of the finest music ever

---

*In fact, goose liver pâté is basically equal parts goose liver and butter, with some wine and spices.

composed* at a Metro station in Washington, D.C., during the morning rush hour. Would people notice that this guy was better than most buskers? Would they stop to listen? Would they throw a dollar or two his way? Would you?

If you were like 98 percent of the people who passed through L'Enfant Plaza Station that morning, you would have hurried by, oblivious of the performance. Only 27 out of 1,097 (2.5 percent) put money into Bell's open Stradivarius violin case and only 7 (0.5 percent) stopped to listen for more than a minute. Bell played for a little less than an hour and made about $32, which is probably not bad for your basic street performer, but no doubt humbling to a man used to making far, far more for one minute of playing.

Weingarten interviewed a number of people who passed through the station that morning. Of the people who stopped, one recognized Bell from a performance the night before. Another was a serious violinist himself. Another was a Metro worker who, after years of listening to ordinary, albeit occasionally talented, buskers, discerned that Bell was better than average. Aside from these few—and disturbingly to classical music fans, and Bell's fans in particular—people did not stop to listen. Many didn't even look at Bell. When interviewed, passersby said either that they didn't notice the music at all, or that it sounded like a slightly better than average street performer playing everyday classical music. No one expected a world-class musician to be playing technically dazzling pieces in a Metro station. Accordingly, and for the most part, they didn't hear one.

Sometime later I met Joshua Bell and asked him about this

---

*The music included Bach's "Chaconne," Franz Schubert's "Ave Maria," Manuel Ponce's "Estrellita," a piece by Jules Massenet, a Bach gavotte, and a reprise of "Chaconne."

experience. In particular, I wanted to know how he felt about being overlooked and ignored by so many people. He responded that he was really not all that surprised, and admitted that expectation is an important part of the way we experience music. Bell told me that it takes an appropriate setting to help people appreciate a live classical music performance—a listener needs to be sitting in a comfortable, faux velvet seat, and surrounded by the acoustics of a concert hall. And when people adorn themselves in silk, perfume, and cashmere, they seem to appreciate the costly performance much more.

"What if we did the opposite experiment?" I asked. "What if we put a mediocre player in Carnegie Hall with the Berlin Philharmonic? The expectations would be very high but the quality would not. Would people discern the difference and would their pleasure be quashed?" Bell thought for a moment. "In this case," he said, "the expectations would triumph over the experience." Furthermore, he said he could think of a few people who were not great violinists but received wild applause because they were in the right environment.

In the end, I wasn't convinced by Bell's nonchalance about his Metro performance. After all, time heals all wounds, and one of the ways time works in our favor is to help us either forget or misremember the past in a way that makes us feel better about ourselves. Besides, not being surprised that people were too busy to notice his performance must have helped Bell avoid the violinist's version of the old question: "If a tree falls in the forest, and no one is around to hear it, does it make a sound?"

The following day, sitting in the Monterey auditorium, I had the opportunity to listen to Joshua Bell play Bach's famous "Chaconne"—the same wonderful piece that he had played for his commuter audience. I closed my eyes and imagined that, instead of listening to a great violinist, I was hearing a mediocre fifteen-year-old kid play a Stradivarius. I'm no connoisseur, but I swear I could hear a few off-key patches, and some squeaks of the strings suddenly became audible. Perhaps the squeaks were part of Bach's composition, just an inevitable part of playing a stringed instrument, or maybe they were a result of playing in an auditorium rather than a proper concert hall. I could easily imagine how an untrained listener such as myself might attribute these sounds to mistakes of a mediocre player, especially if the player is standing in a bustling train station during rush hour.

At the end of his performance, Bell got a long standing ovation. Though I had enjoyed the performance, I wondered how much the ovation was a reward for his performance and how much was due to the audience's expectations. I'm not questioning the level of Bell's (or anyone's) talent. The point is that we don't really understand the role expectations play in the way we experience and evaluate art, literature, drama, architecture, food, wine—anything, really.

I THINK THAT the role of expectations may have been captured best by one of my favorite authors. In Jerome K. Jerome's 1889 comic novella *Three Men in a Boat*, the narrator and his two traveling companions are at a party at an inn. The discussion happens to turn to comic songs. Two young men, outsiders who lack the aristocratic manners of the other partygoers, assure them that a song by the renowned German comic Herr Slossenn Boschen is the funniest of all, and that Herr Boschen happens to be staying at the very same inn. Perhaps he might be persuaded to play one of his songs for them?

Herr Boschen is quite glad to play for them, and since only the two young men understand German, though everyone else pretends to understand it, the rest of the audience take their cues from them. As the two young men shriek with laughter, so do they. Some members of the audience go a step further and from time to time laugh on their own, pretending to have understood a bit of subtle humor missed by the others.

In reality, it turns out that Herr Boschen is a renowned tragedian and is doing his best to deliver a dramatic, emotionally laden song—while the two young men laugh every few notes in order to fool the rest of the guests into believing that such is the style of German comedy. Confused, Herr Boschen presses on. But when he finishes singing, he leaps up from the piano and pours a stream of German obscenities over his listeners.

Ignorant of German and Germany's musical conventions, the audience members do the next best thing and follow the purported expertise of the two outsiders, laugh on cue from them, and believe that the whole performance, including Boschen's temper tantrum, is uproariously funny. Overall the audience enjoys the performance a great deal.

Jerome's story is exaggerated, but in truth, this is how most of us navigate the world. Across many domains of life, expectations play a huge role in the way we end up experiencing things. Think about the *Mona Lisa*. Why is this portrait so beautiful, and why is the woman's smile mysterious? Can you discern the technique and talent it took for Leonardo da Vinci to create it? For most of us the painting is beautiful, and the smile mysterious, because we are told it is so. In the absence of expertise or perfect information, we look for social cues to help us figure out how much we are, or should be, impressed, and our expectations take care of the rest.

THE BRILLIANT SATIRIST Alexander Pope once wrote: "Blessed is he who expects nothing, for he shall never be disappointed." To me, it seems that Pope's advice is the best way to live an objective life. Clearly, it is also very helpful in eliminating the effects of negative expectations. But what about positive expectations? If I listen to Joshua Bell with no expectations, the experience is not going to be nearly as satisfying or pleasurable as if I listen to him and say to myself, "My god, how lucky I am to be listening to Joshua Bell play live in front of me." My knowledge that Bell is one of the best players in the world contributes immeasurably to my pleasure.

As it turns out, positive expectations allow us to enjoy things more and improve our perception of the world around us. The danger of expecting nothing is that, in the end, it might be all we'll get.

Reflections on Placebos: Don't Take Mine Away! (Chapter 10)

A few years ago, a woman seated next to me on a flight to California took a longish white cylinder from her bag, opened it, and dropped a quarter-size tablet into her airplane cup of water. I watched, mesmerized, as yellowish bubbles fizzed and foamed wildly in the cup. After the activity settled, the woman drank the whole concoction in two large sips.

I was very curious about this and, as she looked very pleased with the whole process, I asked her what she was drinking. She handed me the longish white tube. It was Airborne!

The description on the tube truly impressed me. These

tablets, it said, had the power to boost the immune system and help fight the germs that surround passengers during flights. If I took it at the first sign of cold symptoms or before entering a crowded, potentially germ-infested environment, I could prevent the awful colds that I constantly fought. I could not imagine anything better. And, unlike any other medication I have seen, this one stated clearly that it had been invented by a second-grade teacher! Who better to design cold medications than someone surrounded day in and day out by germ-laden children? Since teachers are continually catching colds from their students, this seemed like a natural connection. Besides, I loved the bubbling, foaming action.

My seatmate could not ignore my enthusiasm, so she asked me if I wanted to try a tablet. I happily accepted one, dissolved it in my half cup of water, watched the fizzing and foaming, then drank the yellowish stuff in one gulp. I could see before me the image of my own beloved second-grade teacher—Rachel—and my fondness for her added to the experience. Almost immediately, I felt better. I completely avoided getting sick after that flight. Proof! Thus did Airborne become a staple in my travels.

Over the next few months I used Airborne as the tube suggested. Sometimes I drank it during a flight, but more often I consumed it after the flight. Each time I repeated the ritual, I immediately felt better about myself and about my chances of fighting off the insidious airborne diseases surrounding me. I was 99 percent sure that Airborne was a placebo, but the bubbles and the ritual were so wonderful that I just knew it would make me feel better. And it did! Besides, taking it made me more confident in my health and less stressed about getting sick—and, after all, stress and anxiety are known to lower immunity.

A few years later, just as I was beginning my book tour and had to fly constantly, I heard the tragic news that Victoria Knight-McDowell, the second-grade teacher from California who invented Airborne, had agreed to pay a sum of $23.3 million in a settlement for false advertising, in addition to refunding money to consumers who bought the product. The manufacturer had to change the statements and claims on the product itself. The former "miracle cold buster" had been demoted to a simple dietary supplement made from 17 vitamins, minerals, and herbs. The old claim that Airborne "supports your immune system" remained intact on the packaging, but was accompanied by one of those pesky daggers (†) indicating fine print. You have to search for it, but eventually you find it hidden away in the back corner: "These statements have not been evaluated by the Food and Drug Administration. This product is not intended to diagnose, treat, cure, or prevent any diseases." How depressing.*

So there I was, faced with at least three flights a week for the next few months, and the magic of my Airborne was ripped away from me. I felt as if I had learned that a person I'd considered a good friend for many years had never really liked me and had been saying bad things about me behind my back. Maybe, I thought, if I went straight to the drugstore and got some of the old containers with the exaggerated and inflated claims, they might help restore the magical power of

---

*I suspect that Airborne incorporates many elements to maximize the placebo effect (bubbles, foaming, medicinal color, exaggerated claims, and so on) and, as a consequence, had a real beneficial impact on my immune system and my ability to fight off illnesses. Placebos are all about self-fulfilling prophecies, and Airborne is one of the best.

Airborne. But this seemed unlikely. I could not avoid the knowledge that my fizzy miracle was no such thing. It was just some dumb vitamin with neat Alka-Seltzer special effects. In the face of such disillusionment, I can no longer enjoy the wonderful placebo-immunity-enhancing effect of yesteryear.

Oh why, why did they do this to me? Why did they take my wonderful placebo away?

Thoughts about the Subprime Mortgage Crisis and Its Consequences

For a long time, economists have maintained that human behavior and the functioning of our institutions are best described by the rational economic model, which basically holds that man is self-interested, calculating, and able to perfectly weigh the costs and benefits in every decision in order to optimize the outcome.

But in the wake of a number of financial crises, from the dot-com implosion of 2000 to the subprime mortgage crisis of 2008 and the financial meltdown that followed, we were rudely awakened to the reality that psychology and irrational behavior play a much larger role in the economy's functioning than rational economists (and the rest of us) had been willing to admit.

It all started from questionable mortgage practices, augmented by collateralized debt obligations (CDOs are securities based mostly on mortgage payments). In turn, the CDO crisis accelerated the deflation of the housing market bubble, creating a reinforcing cycle of decreasing valuations. It also brought to light some questionable practices of various players in the financial services industry.

In March 2008, JP Morgan Chase acquired Bear Stearns at two dollars per share, the low valuation resulting from the fact that Bear Stearns was under investigation for CDO-related fraud. On July 17, major banks and financial institutions that had bet heavily on CDOs and other mortgage-backed securities posted a loss of almost $500 billion. Eventually 26 banks and financial institutions would be under investigation for questionable practices relating to their handling of CDOs.

On September 7, the government federalized Fannie Mae and Freddie Mac to avoid their bankruptcy, which would have had dire effects on financial markets. A week later, on September 14, Merrill Lynch was sold to Bank of America. The following day, Lehman Brothers filed for bankruptcy, raising fears of a liquidity crisis that could precipitate an economic meltdown. The day after that (September 16) the United States Federal Reserve lent money to the insurance giant AIG to prevent the company's collapse. On September 25, after being seized by the Federal Depositor Insurance Corporation (FDIC), Washington Mutual was forced to sell its banking subsidiaries to JP Morgan Chase, and the following day the bank's holding company and remaining subsidiary filed for Chapter 11 bankruptcy.

On Monday, September 29, Congress rejected the bailout package (formally known as the Troubled Assets Relief Program, or TARP) proposed by President Bush; resulting in a 778-point drop in the stock market. And while the government worked to build a package that would pass about a week later, Wachovia became another casualty as it entered talks with Citigroup and Wells Fargo (the latter eventually bought the bank), and the stock market reacted to the news of the bailout with a loss of 22 percent of its valuation, making this the worst week on Wall Street since the Great Depression.

One by one, the institutional banks—all staffed by wonderful (rational) smart economists, who followed standard models, fell like so many dominoes.

If the rational economic approach is not sufficient to protect us, what are we supposed to do? What models should we use? Given our human fallibilities, quirks, and irrational tendencies, it seems to me that our models of behavior and, more important, our recommendations for new policies and practices should be based on what people actually do rather than what they are supposed to be doing under the assumption that they are completely rational.

This seemingly radical idea is, in fact, a very old idea in economics. Before Adam Smith, the grandfather of modern economics, wrote his magnum opus, *Inquiry into the Nature and Causes of the Wealth of Nations* (1776), he wrote *The Theory of Moral Sentiments* (1759), a book that is equally important but much more psychologically oriented. In *The Theory of Moral Sentiments*, Smith notes that emotions, feelings, and morality are aspects of human behavior which the economist should not ignore (or, worse, deny) but instead treat as topics worthy of investigation.

About 200 years ago, another economist, John Maurice Clark, noted similarly, "The economist may attempt to ignore psychology, but it is sheer impossibility for him to ignore human nature. . . . If the economist borrows his conception of man from the psychologist, his constructive work may have some chance of remaining purely economic in character. But if he does not, he will not thereby avoid psychology. Rather, he will force himself to make his own, and it will be bad psychology."[31]

How did economics move from embracing human psychology to completely rejecting the possibility that human behavior could be irrational? One reason, no doubt, has to do with the fascination economists have with simple mathematical models. Another has to do with their desire to provide businesses and policy makers with simple, tractable answers. And while both of these can be good reasons to sometimes ignore irrationality, they also take us down a dangerous road.

In my mind, the goal of behavioral economics is to rekindle the economic interest in human behavior and psychology that Adam Smith wrote about. In general, researchers in behavioral economics are interested in modifying standard economics so as to take real, common, and often irrational behavior into account. We want to move the study of economics away from being grounded in naive psychology (which often fails the tests of reason, introspection, and—most important—empirical scrutiny), and return it to a more broadly encompassing study of human behavior. We think that economics would then become better suited to make recommendations that would help people with their problems in the real world: saving for retirement, educating their children, making decisions about health care, and so on.

IN WHAT FOLLOWS I want to share some of my perspectives, from a behavioral economics angle, about this strange new world to which we've all so suddenly awakened. What brought us to our present economic mess? How can we better understand what happened? How might we start thinking about our next steps, to make sure we are not going to get into such deep trouble again? The answers to the questions below are not based on experiments with the stock market itself, because the nature of the stock market makes it very hard to conduct any direct experiments. Instead, they are based on general experimental findings in psychology, economics, and behavioral economics, offered from my personal and professional perch, and they should be taken with an appropriate amount of salt.

(1) Why did people take on mortgages that they couldn't really afford?

Politicians, economists, newscasters, and the public have placed the blame for large and risky mortgages on different parties. Some think irresponsible borrowers assumed more debt than they had reason to believe they could afford. Others think borrowers only followed the guidance of predatory lenders, who at the time were thought to be experts. It seems to me that both accounts have some truth to them, but I also think that the main culprit is the inherent difficulty of figuring out the ideal amount of mortgage someone in a particular financial situation should take on.

Here is the crux of the problem: When the housing market was hot, the bankers who gave out mortgages assumed, logically, that the customers would not want their houses to go into foreclosure. To further ensure that people would repay their loans, the mortgage contracts also included a variety of penalties and fines, in case people decided to walk out on their mortgages. On first glance this logic seemed very appealing: given all the terrible things that could happen to those unable to repay a loan (loss of their homes, wrecked credit, foreclosure fees of different sorts, legal fees, and the possibility of being sued by the lender for a deficiency), the banks assumed that people would try very hard not to over-borrow.

Think about it this way: Imagine that I have agreed to lend you as much money as you want and have promised you that I will break both of your legs if you fail to return the money to me. Under these conditions, wouldn't you try very hard not to borrow too much money and work hard to repay me on time? But as anyone who's ever watched a Mafia movie knows, something always goes wrong with such deals. What once seemed like a logical process often ends up depending on highly questionable assumptions. In the leg-breaking scenario, the assumption is that you can figure out the amount you can repay without risking your legs. And in the mortgage scenario, the central assumption is that people can figure out the optimal amount they can borrow without risking their homes. Of course, with a mortgage the computation is more complex, as it needs to take into account taxes, inflation, changes in property values, and more.

As in the story of Goldilocks and the three bears, there's a loan amount that is just right—it's not too small and it's not too large. But can people really compute the "just right" amount that they should borrow?

DURING THE TIME when the housing market was heating up (more or less between 1998 and 2007), I had the privilege of having an office at the research department of the Federal Reserve Bank in Boston. I would spend most of my time at MIT, but once a week I would show up at "the bank," where my job was to debate with the economists on staff and try to inject some awareness of behavioral economics into their work. One day I got into a discussion with one of the local economists—let's call him Dave—about the limits that banks and the regulators should place on mortgages. Dave advocated eliminating all barriers from the mortgage process. He thought that all home buyers were perfectly able to make the optimal decision for their particular circumstances.

Sumi and I had moved into our new house a few months earlier. Having just gone through the mortgage process myself, I had a different perspective. When we were trying to figure out how much to spend on a house, I asked some experts I knew—including a few finance professors at MIT and some investment bankers—what seemed like two simple questions: (1) Given our financial situation, how much should we spend on a house? and (2) How much should we borrow on a 30-year mortgage?

Everyone I asked told me the same thing—that our monthly payments on a mortgage should come to no more than 38 percent of our joint monthly income, and that this amount, together with the interest rate, would dictate the maximum that we could borrow. But this did not answer the question I had asked, and when I tried to push for an answer, the experts told me that they had no way to help me figure out the ideal amount we should spend or borrow. I repeated this story to Dave but he quickly dismissed my concerns. He informed me that even though no one can figure out the optimal amount to borrow, everyone can figure out the general amount, and the small mistakes people make here and there don't really matter much.

I wasn't entirely comfortable with this blanket approach, so I decided to conduct a small study to examine how people actually determine how much to borrow. The housing market was in full bloom, and there was no problem finding people who were house hunting and willing to share their thoughts and decision processes with me. What I found was that the average home buyer (that is, everyone I interviewed aside from Dave) really does have a hard time figuring out how much they should borrow. So, instead of figuring out the answer to the correct question (how much should we borrow?), they focus on a different question altogether, one that is not the correct question, but one that they can easily answer: how much *can* we borrow? They use a mortgage calculator, talk to an enthusiastic realtor or two, and figure the maximum payment that they can make every month, which, on average, is roughly 38 percent of their income. From there they figure the maximum mortgage that a bank will lend them, and this determines the price of a house that they end up looking for and buying.

This story of how people figure out their mortgage offers a general lesson in human decision making. When we can't figure out the right answer to the question facing us, we often figure out the answer to a slightly different question, and apply this answer to the original problem. This is how a question about the optimal amount to borrow transforms itself into one about the largest amount that a bank is willing to lend. But these are not equivalent questions at all.

Think about it. If you had to buy a new house right now, what is the ideal amount you should spend? How much of it should you take as a mortgage? If you can't figure this out on your own, and the bank and all the mortgage calculators tell you that you can borrow up to 38 percent of your salary to cover monthly payments, wouldn't you accept this amount as an implicit recommendation for how much you should be borrowing?

A FEW WEEKS after our discussion, Dave was assigned to write a paper on the wisdom of interest-only mortgages.* He was very excited about these types of mortgages and wanted to recommend that the regulators promote them as much as possible. "Look," he tried to explain to me, "there is no argument that interest-only mortgages are more flexible than regular mortgages. The people who take them on can decide each month what they want to do with the money that otherwise, in a regular mortgage, would go toward paying the principal. They can pay down their credit card debt or pay for college tuition or medical expenses. Or, if they prefer, they can always pay down the principal of their mortgage."

I nodded, waiting for him to unspool. "Go on," I said.

"So at a minimum, interest-only mortgages are as good as regular mortgages. But they give people more flexibility in terms of how they want to spend their money, and by definition every time you add flexibility you help consumers because you increase their freedom to make the decisions that are right specifically for them."

I said that it all sounded perfectly reasonable, assuming that people make perfectly reasonable decisions. Then I told him about the results of my small study, which had left me uncomfortable. "If people simply borrow the most they can," I explained, "interest-only mortgages will not increase the fi-

---

*An interest-only mortgage is a loan that works as follows: over the life of the loan, the borrower is required to pay only the interest, and as a result, at the end of the loan period the balance is the same as the initial loan. For example, if you took a 10-year loan of $300,000 at a 6.25 percent interest rate, a regular mortgage would cost you $3,368.40 a month, whereas an interest-only mortgage for the same amount and at the same interest rate would cost you only $1,562.50 a month. Of course, if you took the regular mortgage, you would owe nothing by the end of the 10 years and would also own your home, but if you took the interest-only mortgage, you would still owe $300,000 (at which point you will take on a new mortgage, and so on).

nancial flexibility of the people who use them. They will only increase the amount that people will borrow."

Dave was not persuaded, so I tried to give him a more concrete example. "Let's say your cousin what's-her-name . . ."

"Didi," he volunteered.

"Let's say that Didi can afford a regular mortgage with a monthly payment of three thousand dollars. Now you give her the option of taking an interest-only mortgage. What will she do? She could of course get a house she can afford with the regular mortgage and simply pay less every month—using the extra money to pay her student loans. But if she's like other people, Didi will use her maximum ability to pay as the starting point for figuring out what mortgage and house to get, and she'll end up paying three thousand dollars a month for a much bigger, fancier home. She will not have any more flexibility, but she will be much more exposed to the housing market."

I don't think Dave was very impressed with my arguments. But after the subprime mortgage crisis hit, I had the opportunity to look at some data on interest-only mortgages, and it did appear that instead of providing financial flexibility, all that they achieved was to stretch borrowing and put borrowers at higher risk in a fickle housing market.

FROM MY PERSPECTIVE, one of the main failures of the mortgage market was that the bankers didn't even consider the possibility that people cannot compute the right amount to borrow. If banks understood this, they undoubtedly would not have left it up to individuals to figure out the right amount for themselves. In the absence of such an understanding, however, the banks tempted individuals to borrow more than they could possibly afford. Sure, the banks could threaten borrowers with the financial equivalent of breaking legs, but they didn't help borrowers do what's best for the banks—or themselves. It's no wonder that, when the housing crisis finally hit, both banks and their customers wound up with broken legs.

Now, let's say that after all is said and done, the banks finally wise up and decide to conduct empirical studies that examine how people might go about computing the ideal loan amount. Assuming their data reveal the same results as my small study (that people simply borrow to the maximum), the bankers might then realize that it is in their best interest to help borrowers make better decisions. How could they do this?

Obviously, helping borrowers figure out a realistic mortgage amount is not going to be simple, but I know we can do much better than what mortgage calculators do right now (in fact, I don't think we could do worse). So let's say the banks accept the challenge and actually develop better mortgage calculators that not only tell people the maximum they could theoretically borrow, but also help people figure out the right amount for them. If people had the benefit of such humane mortgage calculators, I suspect that they would make better decisions, take on less risk, and ultimately be less likely to lose their homes. Who knows? If such calculators had existed during the last 10 years, maybe much of the mortgage fiasco would have been avoided.

Despite my belief in the desire of borrowers to make the right decisions (and to avoid the disastrous outcomes of making wrong decisions), I must admit that even if some of the banks had created better mortgage calculators, it is possible that in the delirium of the housing market bubble, zealous bankers and real estate brokers could still have pushed people to borrow more and more.

This is where regulators could have stepped in. After all, regulation is a very useful tool to help us fight our own worst tendencies. In the 1970s, regulators placed strict limits on mortgages. They dictated the share of income that could be used to pay a mortgage, the amount of down payment required, and the proof that borrowers had to show to document their income. Over time, these limits were dramatically and dangerously relaxed. Eventually, banks offered the infamous NINJA mortgages ("No Income, No Job or Assets") to people who should never have taken out loans in the first place, and thus ushered in the subprime mortgage fiasco.

You see, in a perfectly rational world, it would make sense to eliminate all limits and regulators from all markets, including the mortgage market. But because we don't live in a perfectly rational world, and because human beings don't always naturally make the right decisions, it makes sense to limit our ability to cause damage to ourselves and others. This is the real role of regulations—they provide us with safe boundaries. They limit our ability to drink and drive; they force kids to go to school; they make pharmaceutical companies empirically test the medications they administer; they limit the ability of companies to pollute our environment; and so on. Certainly, there are many domains of life in which we can function reasonably well without regulations, or at least not cause too much damage when we are left to our own devices. But, when our ability to perform at a satisfactory level is low or nonexistent, and when our failures can hurt ourselves and others (think about driving)—this is when regulations are very handy boundaries to apply.

(2) What caused bankers to lose sight of the economy?

The financial crisis of 2008 left a lot of people feeling that the investment bankers involved were fundamentally evil human beings, and that the economic crisis resulted from their deceitfulness and greed. Certainly, people like Bernard Madoff were out to cheat their investors for personal gain. But personally, I think calculated cheating was the exception rather than the rule in this financial fiasco.

I'm not suggesting in any way that the bankers were innocent bystanders, but I do think that the story of their actions is more complicated than simply accusing them of being bad apples. As in the aftermath of the Enron case and other market failures, it is important to understand what caused the bankers to behave as they did, since this is the only way to ensure that we don't repeat these same mistakes. To that end, let's take stock of what we know about conflicts of interests—a very common foible in the modern workplace.

THE "THEORY OF RATIONAL CRIME" was, perhaps not surprisingly, born in Chicago—a city known for shady politicians, organized criminals, and rational economists. There the Nobel Prize–winning economist Gary Becker first suggested that people who committed crimes applied rational analyses of opportunities and costs. As Tim Harford describes in his book *The Logic of Life*, the birth of Becker's idea was quite mundane. Becker was late for a meeting, and legal parking was scarce, so he decided to park illegally and risk getting a ticket. Becker contemplated his own thought process and behavior in this situation and realized that in planning this crime, there was no place for morality. It was entirely a matter of expected costs and benefits. He weighed the chance that he would be caught and the cost of being fined against the difficulty of finding a legal spot and getting to the meeting even later. He decided to risk the parking ticket, and performed the only crime suited for an econo-mist—a perfectly rational crime.

According to the theory of rational crime, we all should behave like Becker. This means that the average person, like your average mugger, merely makes his way through the world serving his own interests. Whether we do this by rob-bing people or writing books is inconsequential. What is im-portant is how much money is at stake, the likelihood of getting caught, and the magnitude of the expected punish-ment. It's all about weighing our costs and benefits.

THIS RATIONAL COST-BENEFIT approach to decisions in gen-eral, and to crime in particular, may describe Gary Becker himself very accurately—but as we saw in Chapters 11 and 12, simple cost-benefit calculations do not seem to capture the real forces that drive most of us to cheat or to behave honestly. Instead, the picture that emerges from our experi-ments is that cheating arises from our attempts to balance two incompatible goals. On one hand, we want to look in the mirror and feel good about ourselves (ergo, "I can't even look at myself in the mirror" is an indicator of one's own guilt). On the other hand, we're selfish, and we want to benefit from cheating. On the surface, these two motivations seem contra-dictory, but our flexible psychology allows us to act on both of them when we cheat "just by a bit"—benefiting financially from cheating while at the same time managing not to feel bad about ourselves. I think of this as an individual "fudge factor," or a fuzzy conscience.

One way to look at the experiments described in Chapters 11 and 12 is to think about them as an examination of what happens when people wrestle with conflicting interests. When we placed participants in situations in which they were torn between wanting to behave honorably and wanting to benefit financially, they usually succumbed to temptation but only by a little bit. In that light, consider the situation facing a physician who has been involved in research with a pharmaceutical firm, and who gets profit sharing from the company's new prescription drug for, say, diabetes. In treating a patient with diabetes, the doctor can choose a standard drug that he knows will work well. But, he can also write a prescription for the new drug, in which he has a financial interest. He suspects that the standard drug is slightly better for the patient, but the new treatment would benefit his practice. If the diagnosis is very clear-cut, the physician would most likely recommend the best treatment for the patient. But if there is some uncertainty, as is the case in most medical decisions, the physician would most likely recommend the drug he helped develop, allowing himself to both feel good about his diagnosis and at the same time benefit financially from it.

Such conflicts of interests are not limited to doctors, of course; they appear in every aspect of life. Take sports, for example—if you are a fan of a particular sports team and the referee makes a close call against your team, you will most likely think of the referee as blind, idiotic, or evil. Managing to see reality from a self-serving perspective is not an exclusive moral flaw, limited only to "bad people." It's a common human foible and is part of being human. As we discussed in Chapter 9, when we expect something, we are likely to repaint reality in the colors that we want to see. We filter information through our eyes and brains in accordance with our expectations and patterns—and we are very good at convincing ourselves that what we wanted to see is indeed what we saw.

SEEN THROUGH THE lens of conflict of interests, some aspects of the 2008 financial crisis become clearer. It seems to me that with few exceptions, bankers wanted to accurately estimate the risks associated with different financial products, and make good investments for themselves and for their clients. On the other hand, they also had tremendous financial incentives to view financial products such as mortgage-backed securities as fabulous innovations. Put yourself in their shoes: If you could make $10 million simply by getting all your clients to buy mortgage-backed securities, wouldn't you soon convince yourself that such investments were truly wonderful? And if you had to buy into the story of rational markets to convince yourself that this was the case, wouldn't you become a true believer? As with sports fans, bankers' conflicts of interest gave them a reason to see the market make calls in their favor, and thanks to their ability to observe the world as they expected, they managed to see mortgage-backed securities as the best human invention since sliced bread.

On top of the basic conflict of interest, the bankers had one more force working against them—the power of fuzziness. As I described in Chapter 12, when the participants in our studies had an opportunity to cheat for tokens that were one step removed from money for a few seconds, they doubled their cheating. In the same way that the tokens in our experiments allowed our participants to bend reality, I suspect that the opaque nature of pricing mortgage-backed securities, derivatives, and other complex financial products allowed bankers to see what they wanted to see, and to be dishonest to a larger degree. When it came to these complex financial tools, conflicts of interest caused the Wall Street giants to want to see them as the latest and greatest innovations of the modern world, and thanks to the inherent fuzziness of these financial instruments, they were more easily able to reshape reality in a way that was comfortable for them.

So there they were, and here we are. In a market driven by the all-too-human desire to prosper, our incredible ability to fudge reality and shape it to suit our vision got us into trouble. The stock market also utilized lots of fuzzy signifiers for money. For example, bankers often use the term "yard" to signify a billion, "stick" to mean a million, and "points" to mean hundredths of a percent—tokens on a grand scale. Together all these factors allowed the bankers' natural ability at bending reality to flourish and led to new levels of deception.

OF COURSE, THERE'S the ultimate question: Where does all this leave us in terms of a solution? If you believe that there are good and bad people, then all you need to do is figure out how to determine who is good and who is bad and hire only the good people. But if you believe, as our results show, that most people faced with a conflict of interest can cheat, then the only solution is to eliminate conflicts of interest.

In the same way that we would never dream of creating a system in which judges get 5 percent of the settlements over which they preside, it's also clear to me that we don't want doctors to sell drugs they help develop, or bankers to be biased by their own incentives. Unless we create a financial system free of conflicts of interest, the sad story of the financial meltdown of 2008 and its terrible aftermath will be repeated.

How do we eliminate conflicts of interest from the markets? We can hope that the government will begin to regulate the market more effectively, but given the complexity and cost of creating and implementing such regulations, I'm personally not holding my breath until this solution is in place. My hope is that one of the banks will decide to step up to the challenge and lead the way for others by announcing a different pay structure, different incentives for its bankers, transparency, and strict rules against conflicts of interest. I also think that such actions will eventually be beneficial for the bank.

While I wait for an upstanding bank and better regulations, I plan to take a proactive step by looking more closely at my relationships with physicians, lawyers, bankers, accountants, financial advisers, and the other professionals to whom I turn to for expert advice. I can ask doctors who prescribe me drugs whether they have any financial interest in the pharmaceutical company; financial advisers whether they get paid by the management of particular funds they are recommending; and life insurance salespeople what kind of commission they are working on—and seek to establish relationships with providers who do not have conflicts of interest (or at least get a second independent opinion).

While I realize that doing so will be time-consuming and expensive, I suspect that acting on the biased opinions from a specialist with a strong conflict of interest could end up costing me more in the long run.

(3) Why didn't we plan better for the possibility of bad times?

The general phenomenon social scientists call the planning fallacy has to do with our tendency to underestimate how long we will take to finish a task (it explains why roadwork never seems to get finished and new buildings never open on time). There is a very simple way to demonstrate the planning fallacy. Ask some undergraduate students how long it will take them to finish a big task, such as their honors thesis, under the best conditions.

"Three months" is the standard reply.

Next ask them how long it would take under the *worst* conditions.

"Six months," they routinely offer.

Then ask another group how much time they think it will really take them to finish their honors thesis under *normal* conditions, following their usual study, work, and activity schedule.

"Three months," they usually respond.

Given the first two answers, you might expect that they would predict that finishing their honors thesis would take them closer to six months, or maybe four and a half months, but they don't. Their answer is always too optimistic no matter how unrealistic this may actually be. If you think that this kind of misjudgment occurs only with undergraduates, think back to the times you promised your spouse that you would be home from work by six p.m. You have every intention of fulfilling your promise, but invariably something goes wrong and your departure is delayed. You get a call from a client, you receive an e-mail from your boss that needs an immediate response, a coworker stops by your office to sound off about something, or perhaps you've been trying to print something out and the printer gets stuck. Now, if the printer got stuck for five minutes every time you tried to print, you would quickly take this into account and plan to print before you needed to depart the office. But because different things go wrong at different times, and because it's impossible to predict when any particular delay will rear its ugly head, we play the scenario of leaving the office in our minds (sending one last e-mail, printing the notes for tomorrow's meeting, packing our bag, finding our keys, and leaving for the day), without taking any of these possible interruptions or mishaps into account.

As it turns out, the planning fallacy also plays a large role in how we think about our budgets. When we think generally about what we can and cannot afford, and what we should and should not buy, we consider our monthly bills and expenses and make decisions more or less accordingly. But when things go awry and something unplanned happens—say we need a new roof for our house or a new set of tires for our car—we just don't have the cash in the bank to pay for them. Because different bad things happen at different times, we don't take many of them into account.

REGRETTABLY, THIS IS not the end of the story, because the planning fallacy joins forces with its silent partner the financial industry, and together they wreak even more havoc on our lives. It turns out that the financial industry understands that we are partially blind to these negative events, and this is exactly where the industry sticks it to us. When something goes badly and we don't make a payment on time or bounce a check, there are strong negative consequences. To illustrate, allow me to tell you a story about my experience of being poor for a day, and what I learned in the process.

In the winter of 2006 I was out of the country for a month, and during that time my car insurance expired. When I returned and discovered this, I called my insurance agent to request a renewal of my policy. "No, no, no," she said with surprising vehemence. "If your insurance has expired, you can't renew it over the phone. You have to come to our office in person and take out a new policy."

I was living in Princeton, New Jersey, at the time, and my insurance agency was about 250 miles away, in Boston. I tried to argue with the agent and even called a few other insurance agencies, but they all had the same demand. Because I had let my insurance expire, I was now categorized as a bad person in the eyes of their industry, and an agent had to see me face-to-face to approve a policy. So I took the seven-hour train trip to Boston and arrived at the insurance office in the early afternoon—ready to hand over a check, renew my insurance, and then take the train back to Princeton. You would think that the rest would be simple, but of course, it wasn't.

The first thing I learned when I got to the insurance agency was that my insurance premium would increase substantially. Sheila, the insurance agent, informed me that in allowing my insurance to lapse, I lost all the good-driver discounts I'd collected. Now that I was a subprime human being, I was given a premium suitable for a teenager. Furthermore, the insurance company would not accept a check from me, because, in their eyes, I had shown my true, irresponsible colors.

"Will a credit card do?" I asked, in as even a tone as I could muster.

"Of course not," Sheila said coldly. Her hands were hidden under her desk. I imagined that at any moment she might push a button to call the police. "We can accept only cash from you."

For some reason I had a few hundred dollars in cash on me, and there was a bank just next door. Using two ATM cards to withdraw all the cash I could, I was able take out an additional $800 for a net sum of slightly more than a half year's worth of insurance.

"Surely this will do?" I said, plunking down the money in front of Sheila. "I am giving you the payment for the first six months of coverage now in cash, and I'll send you the rest tomorrow."

Sheila paused and looked at me as if I were hard of understanding (which I guess I was). "You must pay us the premium for a year in cash," she said very slowly, "before we can renew your insurance." Then her face suddenly brightened. "Luckily," she added in a more cheerful tone, "we have a solution designed for this exact problem. There is a lending company that offers short-term loans just for these cases. The application process is very quick and simple and you can be out of here in ten minutes."

What else could I do? I asked her to sign me up for this special loan. The terms of the loan included a 20.5 percent interest rate on the loan itself plus a $100 fee just for the privilege of enrolling. This was obviously very annoying, but I had no other option if I was to get my insurance back that day. (Of course, a few days later I paid this terrible loan off.)

On the train back to Princeton, I concluded that this had been a maddening but very enlightening experience. I learned that the moment you make a financial mistake, the chances are very high that you will be hit with all kinds of fines, bureaucratic difficulties, and additional financial obstacles. I was fortunate not to have suffered much: Sure, I'd lost a day of work, and I had been forced to pay money for the train tickets, the fee to initiate the loan, the expensive loan itself, and the increased premium on the insurance—in advance and in cash. But what, I wondered, happens to people who can't afford to take a day off work, and who are on the verge of financial difficulty? How do they come up with the money to pay all these fees and higher premiums? If I were stretched to my limit with financial obligations, and if I had no cushion, this incident would have most likely pushed me over the financial edge, making my life much more expensive, stressful, and difficult. I would have had to take out a terribly expensive loan to pay for my car insurance, borrow more to pay that loan, start carrying a balance on my credit card, start paying fees and high interest for that privilege, and so on.

I later learned that many parts of the insurance and banking industry operate in such a way as to take advantage of people who are already at financial risk. Think, for example, about the "perk" of free checking that the banks so generously provide us. You might think that banks lose money by offering free checking, because it costs them something to manage the accounts. Actually, they make huge amounts of money on mistakes: charging very high penalties for bounced checks, overdrafts, and debit card charges that exceed the amount in our checking accounts. In essence, the banks use these penalties to subsidize the "free checking" for the people who have sufficient amounts of cash in their checking accounts and who are not as likely to bounce a check or overdraw with their debit cards. In other words, those living from paycheck to paycheck end up subsidizing the system for everyone else: the poor pay for the wealthy, and the banks make billions in the process.

Nor does the usury (I daresay, depravity) of the banks end

there. Imagine that it is the last day of the month and you have $20 in your checking account. Your $2,000 salary will be automatically deposited into your bank today. You walk down the street and buy yourself a $2.95 ice cream cone. Later you also buy yourself a copy of *Predictably Irrational* for $27.99, and an hour later you treat yourself to a $2.50 caffe latte. You pay for everything with a debit card, and you feel good about the day—it is payday, after all.

That night, sometime after midnight, the bank settles your account for the day. Instead of first depositing your salary and *then* charging you for the three purchases, the bank does the opposite and you are hit with overdraft fees. You would think this would be enough punishment, but the banks are even more nefarious. They use an algorithm that charges you for the most expensive item (the book) first. *Boom*—you are over your available cash and are charged a $35 overdraft fee. The ice cream and the latte come next, each with its own $35 overdraft fee. A split second later, your salary is deposited and you are back in the black—but $105 poorer.

WE ALL SUFFER from the planning fallacy syndrome, and the banking and insurance institutions, realizing this, build in large penalties that kick in just when these unexpected (un-expected to us) bad things happen. And because when we sign up for these financial or insurance services we certainly don't plan to miss an insurance payment, bounce a check, skip a credit card payment, or go over our debit card limit, we often don't even look at the terms of the penalties, think-ing they do not apply to us. But when "stuff happens," the banks are lying in wait and we end up paying dearly.

Given this modus operandi, is it any wonder that many of the people who took subprime mortgages (by definition those who were not doing well financially) defaulted on their credit card payments, walked out on their mortgages, and even declared bankruptcy?

There are some well-to-do people who think none of this is their problem, of course. But one of the major lessons we've learned from the 2008 economic meltdown is that our financial fortunes are all tied together more tightly than anybody realized. What started as subprime mortgage loans to people with relatively bad credit ended up sucking the wealth out of the entire economy, and bringing almost every economic activity—from car loans to retail spending—to a near-halt. Even people with hefty retirement portfolios took a big hit. In the end, the economy is a complex dynamic system, a bit like the "butterfly effect" in chaos theory where events that happen to a small group of individuals (such as subprime borrowers) can have large and frightening effects down the road for everyone else.

WHAT CAN WE, as individuals, do to overcome the challenges posed by the financial planning fallacy? First, of course, everyone needs to save more for a rainy day* and realize that rainy days are more common than we expect. For people who are already in financial distress, this is clearly not going to be easy to accomplish, nor am I naive enough to think that we can completely eliminate the problem of the financial planning fallacy. But we can guard ourselves against the bumps in the financial road by putting aside some money to

---

*For a helpful perspective, see M. P. Dunleavy, "Making Frugality a Habit," *New York Times* (January 9, 2009).

give us a cushion, and by so doing we might be able to put a dent in the planning-fallacy problem and make it less acute.

Finally, I think that punitive finance practices—including high-interest credit cards, car title loans, payday loans,* and the like—that prey upon those with the fewest resources have to be controlled. It is more appropriate, fair, and better for the economy, as a whole, if we spread the cost of financial services such as checking accounts, credit cards, and insurance among all customers rather than forcing those with fewer resources and fewer options to carry a large part of the burden. At the end of the day we have to realize that when we financially squeeze people who don't have much financial juice in them, it hurts all of us.

(4) Did the government overlook trust as an important economic asset?

In September 2008 Henry Paulson, who was then secretary of the Treasury, told American legislators and the public that unless they immediately coughed up a substantial amount of money ($700 billion) to buy toxic securities from the banks, devastation would result. When this bailout plan was proposed, it looked as if the American public really wanted to strangle the bankers who had flushed our portfolios down the toilet. (The eventual name of the bailout package was "The Troubled Asset Relief Program," but this did little to change the sentiment on the street.)

One nearly apoplectic friend of mine promoted the idea of "an old-fashioned, 1660-style stock market." "Instead of tax-

---

*Since 1990, the number of places in the United States that give "payday" loans has grown faster than the rate at which Starbucks shops have opened.

ing us to bail out those crooks," he ranted, "Congress should put them in wooden stocks, with their feet and hands and heads sticking out. I bet everyone in America would give big bucks for the joy of throwing rotten tomatoes at them!"

My friend was not alone. The following excerpt from a letter by an anonymous lawmaker posted on the politically progressive Web site OpenLeft.com[32] does a good job of describing the public's pent-up rage:

> *I also find myself drawn to provisions (in the bailout bill) that would serve no useful purpose except to insult the industry, like requiring the CEOs, CFOs and the chair of the board of any entity that sells mortgage related securities to the Treasury Department to certify that they have completed an approved course in credit counseling. That is now required of consumers filing bankruptcy to make sure they feel properly humiliated for being head over heels in debt, although most lost control of their finances because of a serious illness in the family. That would just be petty and childish, and completely in character for me. I'm open to other ideas, and I am looking for volunteers who want to hold the sons of bitches so I can beat the crap out of them.*

Rather than taking this anger seriously, and thinking carefully about how to rebuild the public's trust in the banking system and the government, the legislators added insult to injury and contributed further to the erosion of public trust. They added a few irrelevant tax cuts to the proposed bailout package and then force-passed it. A few months later, Paulson revealed that after roughly half of the $700 billion had been paid out to banks, none of the money had been spent to buy back toxic securities, nor did the Treasury Department intend to buy any in the future. He offered no reasons, no explanations, not even so much as an apology. At the end of 2008, when time for bonuses came, the banks did their share to further erode the public's trust by giving themselves millions of dollars in bonuses, and no doubt patting themselves on the backs for a job well done.

To CAST A broader light on the role of trust in society in general, allow me to walk you through an experimental setup that we call the Trust Game. You are one of the two players, and you are paired with an anonymous participant who is your counterpart. The game is played over the Internet, so you will never know each other's identity.

To start, each of you receives $10. You are player 1, and the first move is yours: you must decide whether to keep your money or to send it over to your counterpart. If you keep the money, both of you get to keep your $10 and go home slightly richer. However, if you decide to send the second player your $10, the experimenter quadruples the money you send, so that the other player now has his original $10 plus $40 (the $10 that he got from you, multiplied by four). Now it's his turn to make a decision. He can choose to keep all the money for himself, which means that he would go home with $50 and you would go home with nothing, or he can send half of the money back to you, which means that each of you would go home with $25.

Two questions arise from this basic game: If you are player 2 and your partner has passed you his $10 (earning you an additional $40), would you go home with the $50, or would you share and share alike? And what if you are player 1? In this case you should ask yourself what you expect player 2 to do, and whether he deserves your trust or not. Would you be willing to part with your $10 and risk the possibility that your partner may not share his money with you in turn? The answer to these questions is very simple according to rational economic theory: the second player would never give back the $25, because doing so is not in his financial self-interest. And knowing this, the first player would never give away the original $10 to start with.

So goes the simple, selfish, rational prediction. But think about this for a minute. If you were the second player and the first player sent you $10 (which became $40), would you go home with the $50 or would you send back $25? Would you send your $10? I am not sure what you would do, but it turns out that people in general are more trusting and more recip-rocating than standard economic theory would have us be-lieve. Studies of many versions of the trust game have shown that a significant majority of people send their partner the $10 and that most people reciprocate by sending $25 back.

The Trust Game is very useful as a demonstration of the central role of trust in human behavior, but the story does not end there. A group of Swiss researchers led by a creative and inspiring economist, Ernst Fehr, used an extension of this game to examine the extent of not only trust but also revenge.[33] In the Swiss version of the game, if player 2 decides not to share the $50 with you, you would have a chance to make one more decision. After telling you that the other player has decided to keep the $50, the experimenter would say: "Look, I'm sorry you just lost ten dollars. Tell you what: if you want, you can use some of your own money to wreak a little revenge. For each dollar you give me out of your own money, I'll take two dollars from the other guy. If you give me three dollars I will take six dollars from him; if you give me seven dollars, I will take fourteen dollars from him; and so on. What do you think?"

Once more, put yourself in player 1's shoes. If player 2 betrayed your trust, would you sacrifice your own money to make him suffer?

The experiment showed that most people who had the opportunity to exact revenge on their greedy partners did so, and they punished severely. Yet this finding was not the most interesting part of the study. While the participants were thinking about the prospect of wreaking revenge, their brains were being scanned by a positron emission tomography (PET) machine. And what was the part of the brain that was involved in plotting and executing revenge? It was the striatum, a part of the brain associated with the way we experience reward. This means that a decision to punish untrustworthy partners is related in some way to a feeling of pleasure and reward. What's more, it turned out that those who had a high level of striatum activation punished others to a higher degree. All this suggests that the desire for revenge, even when it costs us something, has biological underpinnings, and that revenge is either pleasurable or somewhat similar to pleasure.

THIS ANALYSIS OF trust and the pleasure of revenge also provides a useful lens through which to view irrationality and behavioral economics more generally. At first glance revenge seems to be an undesirable human motivation: why on earth would human beings have evolved to enjoy seeking revenge on each other? Think about it this way: Imagine that you and I are living 2,000 years ago in an ancient desert land, and I have a mango that you want. You might say to yourself: "Dan Ariely is a perfectly rational person. It took him twenty minutes to find this mango. If I steal it and hide so that Dan will need more time to find me than to go and get a new mango, he will do the correct cost-benefit analysis and set off on his way to find a new mango." But what if you know that I am not rational, and that instead I'm a dark-souled, vengeful type who would chase you to the end of the world, and take back not only my mango but also all of your bananas? Would you still go ahead and steal my mango? My guess is that you would not. In a bizarre way, revenge can be an enforcement mechanism, supporting social cooperation and order.

This is how a human tendency that might initially look senseless, and not part of the basic definition of rationality, can, in fact, be a useful mechanism—not necessarily one that always works in our favor, but one that has some beneficial logic and function nevertheless.

Now that we understand a bit more about trust, its violation, revenge, and mangoes, how do we begin to deal with the current mistrust of the stock market? The parallel between the trust game and the subprime mortgage crisis is very clear: we trusted the bankers with our retirement funds, our savings, and our mortgages, but when it was their turn to act, they walked away with the entire $50 (most likely, you may want to put a few zeros behind that). As a consequence, we feel betrayed and angry, and we want the institutions and bankers to pay dearly.

Beyond the feeling of revenge, this type of analysis helps us understand that trust is an essential part of the economy, and that once trust is eroded, it is very difficult to restore. The central banks can take heroic measures to infuse money, give short-term loans to banks, increase liquidity, buy back mortgage-backed securities, and use any other trick in the book. But unless they rebuild the public's trust, these very expensive measures are unlikely to have the desired effect.

I suspect that the government's actions not only ignored the issue of trust, but they unintentionally contributed to its further erosion. For example, the bailout legislation was eventually passed not because it had been made more appealing, but because a few irrelevant tax cuts were added to it. Also, Paulson asked us to trust him when he said that $700 billion was needed to buy toxic assets, and that he would manage this responsibility appropriately. We learned later that he didn't follow through with the former, and this failure made the latter seem unlikely. And of course, let's not forget the behavior of the bankers themselves, from minor issues such as costly office decorations (Merrill Lynch's CEO John Thain spent more than $1 million to decorate his office), to more substantial issues such as the compensation of the CEOs at Lehman Brothers, Fannie Mae, Freddie Mac, AIG, Wachovia, Merrill Lynch, Washington Mutual, and Bear Stearns—establishing new records for CEO pay.

Imagine how different things would have looked if the banks and the government had understood the importance of trust from the get-go. Had that been the case, they would have worked harder to explain more clearly what went wrong and how the bailout would be used to clean up the mess. They would not have ignored the public's sentiment; they would have used it for guidance. They would have included some trust-building elements in the bailout legislation itself (for example, they could have guaranteed that every bank bailed out with taxpayers' money would have to commit to transparency, limit top managers' salaries, and eliminate conflicts of interest).

Nevertheless, all is not lost. While it is clear that the legislators do not yet understand the importance of trust, I remain hopeful that some of the banks will decide to step away from the herd and be good guys—creating trust by eliminating conflicts of interest and modeling complete transparency. They might do it because it is the moral thing to do or, more likely, because they will understand that the best way to solve the liquidity problem is to engender trust. It will certainly take a while for them to view the world this way, but at some point they will understand that unless they create a new structure to slowly regain our trust, none of us will get out of this economic mess.

(5) What is the psychological fallout from not understanding what the #$%^ is going on in the markets?

At the end of 2008, consumers' confidence was at its lowest since 1967 (the year that research groups began measuring it), suggesting that the economy was also in the worst shape since 1967, and feeding on itself to further sink the economy. While there is no question that the state of the economy was indeed depressing, I suspect that there were other factors—ones not related to the underlying economic situation—that contributed to our gloomy outlook.

Henry Paulson's behavior, as described above, gave us a clear message that no one really understood what was going on in the financial markets and that we had no real control over the monster we had created. One question we might ask is whether the general depression that followed might have been mitigated if Paulson had been able to explain what went wrong in the first place, what his proposed measures were going to achieve, why he changed his decision to buy toxic securities, and what his plan was for the rest of the bailout money.

As it turns out, even some answers could have made a difference. All creatures (including humans) respond negatively in situations where things don't seem to make sense. When the world gives us unpredictable punishments without rhyme or reason, and when we don't have any explanation for what is happening, we become prone to something psychologists call "learned helplessness."

In 1967, two psychologists, Martin Seligman and Steve Maier, conducted a famous set of experiments using one predictable environment, one unpredictable environment, and two dogs—a control dog and an experimental dog.[34] (Warning: The following description may be upsetting for animal lovers.) In the control dog's room, a bell sounded from time to time. Shortly after each bell, the dog received a mild electrical shock—just enough to annoy and surprise him. Fortunately for him, the control dog also had access to a switch that turned off the shocks, and he quickly discovered the switch and learned to use it.

Next door, the experimental dog (which the scientists referred to as the "yoked" dog) received the exact same electrical shocks, but he did not hear any bells sound beforehand. Nor did he have a switch that allowed him to turn off the shocks. From the perspective of their physical reality, both dogs received exactly the same shocks, but the difference in their situations was their ability to predict and control the shocks.

Once the dogs acclimated to their environment (as best they could), the researchers administered a second test. This time, both dogs were put in a "shuttle box"—a large box divided by a low fence into two compartments. From time to time a warning light came on, and a few seconds later the floor of the shuttle box emitted a mild electrical shock. If, at that point, the dog jumped from one compartment to the other, he would escape the shock. Even better, if the dog jumped over the fence to the other compartment when the warning light first came on, he avoided the shock entirely. As you might expect, the control dog quickly learned to jump over the fence as soon as the light went on. Though he was understandably a bit anxious, he seemed relatively happy.

What about the experimental "yoked" dog? You might expect that he would be equally motivated and equally able to escape the shocks in the shuttle box. But the result was both interesting and rather depressing: The yoked dog just lay in the corner of his cage, whimpering. Having learned in the first stage of the experiment that shocks happened unpredictably and inescapably, the yoked dog carried that mindset into the shuttle box. The experience during the first part of the experiment taught this dog that he didn't understand the relationships between cause and effect. As a consequence, this poor dog later became helpless in his general approach to life, exhibiting symptoms similar to those of people suffering from chronic clinical depression, including ulcers and a general weakening of the immune system.

You MIGHT THINK that this experiment applies only to electrical shocks and dogs, but the principle holds true in many cases when we don't understand the causes of rewards and punishments in our environment. Imagine yourself in the economic equivalent of the yoked dog's chamber. One day you are told that the best place to invest your money is in high-tech stocks, and the next moment, without warning— *bzzz*—the Internet stock bubble explodes. Next, you are told that the best place to invest your money is housing, and again—no warning, then *bzzz*—the value of your house plummets. Then, suddenly—*bzzz*—the price of gasoline increases to an all-time high, presumably due to the war in Iraq, yet a few months later, even as the war rages on, the price of gasoline drops—*bzzz*—to a much lower price.

Next you watch as giant financial institutions that have been the backbone of the heretofore-trusted U.S. financial system fail and your investments take a hit—giant *bzzz*. For some unexplained reason some of these institutions receive a bailout—*bzzz*—using the money you earned and then paid in taxes—*bzzz*—and others do not—*bzzz*. Then the Big Three automakers find themselves on the verge of bankruptcy (not a real surprise there), but they don't receive the same generous treatment as the banks, even though they were asking for far less and had many jobs on the line. At the end of the day all these dramatically expensive bailout attempts seem like a capricious, idiosyncratic patch-up job with no reason or plan. *BZZZ.*

Does this economic shuttle box sound familiar? All this unexplained and erratic economic behavior destroyed our faith that we understood the causes and effects in our environment and turned the public into the economic equivalent of a yoked dog. As a result of getting zapped with so many different and incomprehensible shocks every now and again, it's no wonder that consumer confidence plummeted and that depression spread.

MEANWHILE, WHAT CAN we personally do to heal from our own learned helplessness? One idea comes from the research of psychologist James Pennebaker at the University of Texas at Austin. Pennebaker's research has repeatedly shown that the active and conscious process of trying to make sense out of difficult, confusing, and even traumatic events can help individuals recover from them. In much of his work, Pennebaker asks his patients to write their reflections in a journal, finding that this helps them a great deal. This means that even when external events make no sense, we can benefit from our own attempts to make sense of our world.

Pennebaker's advice sounds very reasonable, but of course most of us do the exact opposite. We have access to news 24 hours a day on TV, radio, and the Internet—much of it consisting of quick sound bites that aim for our hearts but not our minds. Journalists have a saying: "If it bleeds, it leads," meaning that the top news stories are always the most shocking or sensational ones. It seems to me that most newscasters are shaped by the same mold, with their grave expressions and motionless hair. They also sound as if they've all received standard training in how to come up with quick, sensational sound bites that they repeat every few minutes. Grim stories about the economy take the shape of tear-jerker stories about people who are struggling, who have lost their jobs, and who can't pay for their homes or insurance.

It is not that these stories are unimportant, not very sad, or unuseful, but they don't help us understand what is happening all around us, or what caused the economic meltdown in the first place. And when we submit ourselves to a never-ending daily diet of depressing, emotional sound bites (thinking that we are going to learn something from watching, reading, or listening to them over and over), we risk intensifying our depression. To fight this tendency we should follow Pennebaker's advice and change the way we consume news from passive receptivity to actively thinking about the information and trying to make sense out of it.

Maybe one day journalists, or Henry Paulson, or the next chairman of the Federal Reserve, or Barack Obama, or the new leaders of other government institutions will value our well-being enough to explain to us what is going on and the rationale behind the decisions that they make. And the sooner the better because I am not sure how many more shocks we can take.

(6) Can a global market increase irrational behavior?

For at least the last decade, the globalization of markets has been promoted by many as a good thing. The belief has been that a move from multiple and semi-independent markets toward one big market increases liquidity, encourages financial innovation, and allows friction-free trade. As a consequence, today, in case you haven't noticed, there is not much difference between the Japanese, British, German, and American stock markets. We see them rise and fall almost in unison, if to varying degrees. But as we witness the effects of increased globalization, we should ask ourselves what are the benefits and the costs of having one large market. I suspect that one large market can, in fact, reduce financial innovation, be dangerous to our financial health, and ultimately fail to protect us against financial meltdowns.

To help us think about how one large market can become inefficient, consider the following few paragraphs from *The Lost World* by Michael Crichton.[35] A character named Malcolm (the chaos-theory scientist played in the movie by Jeff Goldblum) goes on a pessimistic rant against cyberspace—pointing out that a world where everyone is connected could bring about the end of creativity, innovation, and evolution.

> *This idea that the whole world is wired together is mass death. Every biologist knows that small groups in isolation evolve fastest. You put a thousand birds on an ocean island and they'll evolve very fast. You put ten thousand on a big continent, and their evolution slows down. Now, for our own species, evolution occurs mostly through our behavior. We innovate new behavior to adapt. And everybody on earth knows that innovation only occurs in small groups. Put three people on a committee and they may get something done. Ten people, and it gets harder. Thirty people, and nothing happens. Thirty million, it becomes impossible. That's the effect of mass media—it keeps anything from happening. Mass media swamps diversity. It makes every place the same—Bangkok or Tokyo or London: there's a McDonald's on one corner, a Benetton on another, a Gap across the road. Regional differences vanish. All differences vanish. In a mass-media world, there's less of everything except the top ten books, records, movies, ideas. People worry about losing species*

diversity in the rain forest. But what about intellec-
tual diversity—our most necessary resource?
That's disappearing faster than trees. But we
haven't figured that out, so now we're planning to
put five billion people together in cyberspace. And
it'll freeze the entire species. Everything will stop
dead in its tracks. Everyone will think the same
thing at the same time. Global uniformity. . . .

Obviously, Malcolm is an intense character with extreme opinions, but even if we don't think that connecting the whole world in cyberspace will bring everything to a halt, it is still interesting to consider whether the connectivity of the global financial markets could actually reduce the diversity of thinking and of financial products, and, as a consequence, decrease competition and efficiency.

Personally, I think that Malcolm's analogy is very apt. I suspect that connecting many markets under the banner of a single global one decreases diversity in financial instruments and in opinions. Moreover, the pressures for conformity are such that living within one global financial village is likely to get all those involved to accept the same general beliefs (model) of how the financial world works. From this perspective, even the rational economic theory would predict that many markets with higher competition between them would be more beneficial than a single market. Somewhat ironically, when rational economic theory has been used to promote one large global market, its supporters have emphasized the benefits of liquidity and efficiency, but they conveniently forget the immense importance of diversity of ideas, approaches, and financial instruments—which at the end of the day is likely to be a more important economic force.

Of course, globalization would be wonderful if the result were a perfect global market. But given the degree to which human beings are prone to mistakes and irrationality, it seems that any market we create is likely to be imperfect. In the end, I would much prefer to have multiple, somewhat independent markets, each perhaps less efficient—but more isolated, flexible, nimble, competitive, and more likely to evolve over time—producing more efficient and robust financial markets.

(7) What is the right amount to pay bankers?

Recently there has been a public outcry against astronomical executive salaries. The basic public sentiment is that it seems unfair that people make so much money for mismanaging our money, especially when it is so difficult to see how bankers' talents and abilities justify their compensation. Naturally, it's particularly offensive when executives receive high bonuses after disastrous performances, or, worse, when the bonuses come from taxpayers' money courtesy of government bailouts.

Not surprisingly, bankers have fought back, claiming that the high salaries are required to attract the best and brightest to crucial, high-stress, high-skill positions, and that the most talented and valuable bankers would go elsewhere if salaries were capped. It is your basic free market argument: if they can't recruit and retain the best minds in business, these minds will simply go elsewhere, leaving us with less qualified people in charge of the economy—and that, in the end, would send us all down the tube.

Rather than seeing this as an ideological debate between self-serving bankers on one side and morally outraged taxpayers on the other, it is more useful to ask what we really know about the relationships between very large bonuses and job performance.

To look at the question of how bonuses affect performance, Uri Gneezy, George Loewenstein, Nina Mazar, and I conducted a few experiments. In one, we gave participants an array of tasks that demanded attention, memory, concentration, and creativity. We asked them, for instance, to fit pieces of a metal puzzle into a plastic frame, to play a memory game that required reproducing a string of numbers, to throw tennis balls at a target, and a few other such tasks. We promised payments of different amounts (either low, medium, or very high bonuses) if they performed any of these tasks exceptionally well. About a third of the subjects were told they'd be given a small bonus (relative to their normal wages), another third were promised a medium-sized bonus, and the last group could earn a very high bonus.

By the way, and before you ask where you can sign up for this experiment, I should tell you that we did the study in India, where the cost of living is relatively low. By doing it there, we could pay people amounts that were substantial to them but still within our research budget. The low bonus was 50 cents, equivalent to what participants could receive for a day's work in rural India. The medium bonus was $5, or about two weeks' pay, and the very high bonus was $50, roughly five months' pay.

What do you think the results were? Would our participants follow the expected reward pattern with the group offered the smallest bonus performing worst, those offered the medium bonus performing better, and those offered the very high bonus performing best? When we posed this question to a group of business students, naturally they expected performance to improve with the amount of the reward. In the business world this assumption is practically a natural law, and the logic that gets executives to command very high pay. But our experiment results revealed the opposite. As it turned out, the group offered the highest bonus did worse than the other two groups in every single task! And the people offered medium bonuses performed no better or worse than those offered low bonuses.

We replicated these results in a study at MIT, where undergraduate students were offered a chance to earn a very high bonus ($600) or a lower one ($60) by performing two four-minute tasks: one that called for some cognitive skill (adding numbers) and another that required only mechanical skill (tapping a keypad as fast as possible). We found that as long as the task involved only mechanical skill, bonuses worked as we usually expect: the higher the pay, the better the performance. But when the task required even rudimentary cognitive skill (as we might suppose investing and banking do), the outcome was the same as in the Indian study: a potential higher bonus led to poorer performance.

Our results led us to conclude that financial rewards are often a two-edged sword. They motivate people to work well, but when these financial rewards get very large they can become counterproductive and actually hurt performance. If our tests mimic the real world, then higher bonuses may not only cost employers more, but also hinder executives in working to the best of their abilities.

INTERESTINGLY, MONEY ISN'T the only thing that compels better (or worse) performance. We conducted a variation of this experiment at the University of Chicago; this time we wanted to look at a different kind of motivator: public image. We asked participants to solve anagram puzzles, sometimes privately in a cubicle and sometimes in front of others. Assuming that their motivation to do well would be higher in public, we wanted to see if being observed by others would affect their performance, and if it would improve or impair their ability. We found that though the subjects did want to perform better when they worked in front of others, they did worse.

We concluded that social pressure, like money, is also a two-edged sword. It motivates people, but having to perform in front of others raises stress too, and at some point that stress overwhelms the benefits of increased motivation.

WHEN I PRESENTED these results to a group of banking executives, they assured me that their own work and that of their employees would not follow the pattern we found in our experiments. (I suggested that with a suitable research budget and their participation, we could examine their assertion, but they were not interested.) I strongly suspect that they were too quick to discount our results. I'd be willing to bet that for the vast majority of bankers, if not for all of them, a multimillion-dollar compensation package could easily be counterproductive because of the stress involved in attaining it, because of the fear of not getting it, and because it takes their minds off the job and focuses their attention on the large bonus.

I don't want to argue that in all situations, regardless of job type or the characteristics of the person, it will be more productive to pay less. But I do want to suggest that compensation is a complex issue involving complex economic incentives, stress, and other aspects of human psychology that we often don't understand and don't take into account. Perhaps the naively simple theory that more money equals better performance is not as practical as we thought, at least not all the time. If more money led to better performance, wouldn't we expect that those who got tens of millions in compensation would be optimal performers? Maybe even perfect? The fact that those with very high salaries and bonuses failed so miserably in the financial fiasco of 2008 should add to the evidence against a direct link between higher rewards and better performances.

IN THE WAKE of an outpouring of public anger, and within weeks of taking office, Barack Obama proposed "common-sense" guidelines for executive pay—at least for companies receiving government money. These measures called for a $500,000 cap on executive salaries; further compensation could be only in the form of stocks, which could not be sold until the government had been repaid. No doubt this makes taxpayers feel better to some extent, but the question is, will it work?

I think not, and here's why: if we were designing the stock market from scratch and offering people $500,000 a year plus stock incentives, I'm sure we would get lots of qualified people who would kill to run a big bank for this compensation. And they might work not only for the salary but also to perform an important civil service in maintaining the financial system on which we all depend. Unfortunately, we're not starting from scratch. Instead, we're dealing with existing bankers who are accustomed to millions of dollars a year in salary, plus millions more in stock options and bonuses. After many years of being conditioned to these circumstances, executives have developed a multitude of reasons why they deserve to be paid so much. After all, how many people do you know who would admit to being paid much more than they're worth?

This is a problem of relativity. The bankers' view of "normal" makes a salary of $500,000 seem both offensive and irresponsible. My guess is that the executives will not accept these conditions; if they do, they'll find other tricks to pay themselves what they think are "right" and fair wages, comparable to what they earned in the past.

If I were Obama's financial czar, I would try to get the bankers, and the system that has given them a warped sense of entitlement, to turn over a new leaf by encouraging the creation of new banks with new pay structures. These new banks would promote the idea that bankers are not greedy bastards but are ethical, upstanding people who fulfill a crucial role that is central for the functioning of the economy and the country (which, in fact, they do). The "old bankers" who feel they needed millions of dollars to do their jobs, and millions more in bonuses to do their jobs well, could try to compete in this new market. But who would want to bank with them when the alternative is a new bank with a more idealistic underpinning and a more realistic, and more transparent, salary structure?

(8) Rational economics has always been the basis for setting up policies and designing our institutions. What's wrong with that?

Neoclassical economics is built on very strong assumptions that, over time, have become "established facts." Most famous among these are that all economic agents (consumers, companies, etc., are fully rational, and that the so-called invisible hand works to create market efficiency). To rational economists, these assumptions seem so basic, logical, and self-evident that they do not need any empirical scrutiny.

Building on these basic assumptions, rational economists make recommendations regarding the ideal way to design health insurance, retirement funds, and operating principles for financial institutions. This is, of course, the source of the basic belief in the wisdom of deregulation: if people always make the right decisions, and if the "invisible hand" and market forces always lead to efficiency, shouldn't we just let go of any regulations and allow the financial markets to operate at their full potential?

On the other hand, scientists in fields ranging from chemistry to physics to psychology are trained to be suspicious of "established facts." In these fields, assumptions and theories are tested empirically and repeatedly. In testing them, scientists have learned over and over that many ideas accepted as true can end up being wrong; this is the natural progression of science. Accordingly, nearly all scientists have a stronger belief in data than in their own theories. If empirical observation is incompatible with a model, the model must be trashed or amended, even if it is conceptually beautiful, logically appealing, or mathematically convenient.

Unfortunately, such healthy scientific skepticism and empiricism have not yet taken hold in rational economics, where initial assumptions about human nature have solidified into dogma. Blind faith in human rationality and the forces of the market would not be so bad if they were limited to a few university professors and the students taking their classes. The real problem, however, is that economists have been very successful in convincing the world, including politicians, businesspeople, and everyday Joes not only that economics has something important to say about how the world around us functions (which it does), but that economics is a sufficient explanation of everything around us (which it is not). In essence, the economic dogma is that once we take rational economics into account, nothing else is needed.

I believe that relying too heavily on our capacity for rationality when we design our policies and institutions, coupled with a belief in the completeness of economics, can lead us to expose ourselves to substantial risks.

HERE'S ONE WAY of thinking about this. Imagine that you're in charge of designing highways, and you plan them under the assumption that all people drive perfectly. What would such rational road designs look like? Certainly, there would be no paved margins on the side of the road. Why would we lay concrete and asphalt on a part of the road where no one is supposed to drive on? Second, we would not have cut lines on the side of the road that make a *brrrrrr* sound when you drive over them, because all people are expected to drive perfectly straight down the middle of the lane. We would also make the width of the lanes much closer to the width of the car, eliminate all speed limits, and fill traffic lanes to 100 percent of their capacity. There is no question that this would be a more rational way to build roads, but is this a system that you would like to drive in? Of course not.

WHEN IT COMES to designing things in our physical world, we all understand how flawed we are and design the physical world around us accordingly. We realize that we can't run very fast or far, so we invent cars and design public transportation. We understand our physical limitations, and we design steps, electric lights, heating, cooling, etc., to overcome these deficiencies. Sure, it would be nice to be able to run very fast, leap tall buildings in a single bound, see in the dark, and adjust to every temperature, but this is not how we are built. So we expend a lot of effort trying to take these limitations into account, and use technologies to overcome them.

What I find amazing is that when it comes to designing the mental and cognitive realm, we somehow assume that human beings are without bounds. We cling to the idea that we are fully rational beings, and that, like mental Supermen, we can figure out anything. Why are we so readily willing to admit to our physical limitations but are unwilling to take our cognitive limitations into account? To start with, our physical limitations stare us in the face all the time; but our cognitive limitations are not as obvious. A second reason is that we have a desire to see ourselves as perfectly capable—an impossibility in the physical domain. And perhaps a final reason why we don't see our cognitive limitations is that maybe we have all bought into standard economics a little too much.

Don't misunderstand me, I value standard economics and I think it provides important and useful insights into human endeavors. But I also think that it is incomplete, and that accepting all economic principles on faith is ill-advised and even dangerous. If we're going to try to understand human behavior and use this knowledge to design the world around us—including institutions such as taxes, education systems, and financial markets—we need to use additional tools and other disciplines, including psychology, sociology, and philosophy. Rational economics is useful, but it offers just one type of input into our understanding of human behavior, and relying on it alone is unlikely to help us maximize our long-term welfare.

IN THE END, I do hope that the debate between standard and behavioral economics will not take the shape of an ideological battle. We would make little progress if the behavioral economists took the position that we have to throw standard economics—invisible hands, trickle-downs, and the rest of it—out with the bathwater. Likewise, it would be a shame if rational economists continue to ignore the accumulating data from research into human behavior and decision making. Instead, I think that we need to approach the big questions of society (such as how to create better educational systems, how to design tax systems, how to model retirement and health-care systems, and how to build a more robust stock market) with the dispassion of science; we should explore different hypotheses and possible mechanisms and submit them to rigorous empirical testing.

For instance, in my ideal world, before implementing any public policy—such as No Child Left Behind or a $130 billion tax rebate or a $700 billion bailout for Wall Street—we would first get a panel of experts from different fields to propose their best educated guess as to what approach would achieve the policy's objectives. Next, instead of implementing the idea proposed by the most vocal or prestigious person in this group, we would conduct a pilot study of the different ideas. Maybe we could take a small state like Rhode Island (or other places interested in participating in such programs) and try a few different approaches for a year or two to see which one works best; we could then confidently adopt the best plan on a large scale. As in all experiments, the volunteering municipalities would end up with some conditions providing worse outcomes than others, but on the plus side there would also be those who would achieve better outcomes, and of course the real benefit of these experiments would be the long-term adoption of better programs for the whole country.

I realize that this is not an elegant solution because conducting rigorous experiments in public policy, in business, or even in our personal lives is not simple, nor will it provide simple answers to all of our problems. But given the complexity of life and the speed at which our world is changing, I don't see any other way to truly learn the best ways to improve our human lot.

FINALLY, I'LL SAY THIS: In my mind there is no question that one of the wonders of the universe is how complex, bizarre, and ever changing human behavior is. If we can learn to embrace the Homer Simpson within us, with all our flaws and inabilities, and take these into account when we design our schools, health plans, stock markets, and everything else in our environment, I am certain that we can create a much better world. This is the real promise of behavioral economics.

Thanks

Over the years I have been fortunate to work on joint research projects with smart, creative, generous individuals. The research described in this book is largely an outcome of their ingenuity and insight. These individuals are not only great researchers, but also close friends. They made this research possible. Any mistakes and omissions in this book are mine. (Short biographies of these wonderful researchers follow.) In addition to those with whom I have collaborated, I also want to thank my psychology and economics colleagues at large. Each idea I ever had, and every paper I ever wrote, was influenced either explicitly or implicitly by their writing, ideas, and creativity. Science advances mainly through a series of small steps based on past research, and I am fortunate to be able to take my own small steps forward from the foundation laid down by these remarkable researchers. At the end of this book, I have included some references for other academic papers related to each of the chapters. These should give the avid reader an enhanced perspective on, and the background and scope of, each topic. (But of course this isn't a complete list.)

Much of the research described in this book was carried out while I was at MIT, and many of the participants and research assistants were MIT students. The results of the experiments highlight their (as well as our own) irrationalities, and sometimes poke fun at them, but this should not be confused with a lack of caring or a lack of admiration. These students are extraordinary in their motivation, love of learning, curiosity, and generous spirit. It has been a privilege to get to know you all—you even made Boston's winters worthwhile!

Figuring out how to write in "non-academese" was not easy, but I got a lot of help along the way. My deepest thanks to Jim Levine, Lindsay Edgecombe, Elizabeth Fisher, and the incredible team at the Levine Greenberg Literary Agency. I am also indebted to Sandy Blakeslee for her insightful advice; and to Jim Bettman, Rebecca Waber, Ania Jakubek, Erin Allingham, Carlie Burck, Bronwyn Fryer, Devra Nelson, Janelle Stanley, Michal Strahilevitz, Ellen Hoffman, and Megan Hogerty for their role in helping me translate some of these ideas into words. Special thanks to my writing partner, Erik Calonius, who contributed greatly to these pages, with many real-world examples and a narrative style that helped me tell this story as well as it could be told. Special thanks also go to my trusting, supporting, and helpful editor at HarperCollins, Claire Wachtel.

I wrote the book while visiting the Institute for Advanced Study at Princeton. I cannot imagine a more ideal environment in which to think and write. I even got to spend some time in the institute's kitchen, learning to chop, bake, sauté, and cook under the supervision of chefs Michel Reymond and Yann Blanchet—I couldn't have asked for a better place to expand my horizons.

Finally, thanks to my lovely wife, Sumi, who has listened to my research stories over and over and over and over. And while I hope you agree that they are somewhat amusing for the first few reads, her patience and willingness to repeatedly lend me her ear merits sainthood. Sumi, tonight I will be home at seven-fifteen at the latest; make it eight o'clock, maybe eight-thirty; I promise.

List of Collaborators

On Amir

On joined MIT as a PhD student a year after me and became "my" first student. As my first student, On had a tremendous role in shaping what I expect from students and how I see the professor-student relationship. In addition to being exceptionally smart, On has an amazing set of skills, and what he does not know he is able to learn within a day or two. It is always exciting to work and spend time with him. On is currently a professor at the University of California at San Diego.

Marco Bertini

When I first met Marco, he was a PhD student at Harvard Business School, and unlike his fellow students he did not see the Charles River as an obstacle he should not cross. Marco is Italian, with a temperament and sense of style to match—an overall great guy you just want to go out for a drink with. Marco is currently a professor at London Business School.

Ziv Carmon

Ziv was one of the main reasons I joined Duke's PhD program, and the years we spent together at Duke justified this decision. Not only did I learn from him a great deal about decision making and how to conduct research; he also became one of my dear friends, and the advice I got from him over the years has repeatedly proved to be invaluable. Ziv is currently a professor at INSEAD's Singapore campus.

Shane Frederick

I met Shane while I was a student at Duke and he was a student at Carnegie Mellon. We had a long discussion about fish over sushi, and this has imprinted on me a lasting love for both. A few years later Shane and I both moved to MIT and had many more opportunities for sushi and lengthy discussions, including the central question of life: "If a bat and a ball cost $1.10 in total, and the bat costs a dollar more than the ball, how much does the ball cost?" Shane is currently a professor at MIT.

James Heyman

James and I spent a year together at Berkeley. He would often come in to discuss some idea, bringing with him some of his recent baking outputs, and this was always a good start for an interesting discussion. Following his life's maxim that money isn't everything, his research focuses on nonfinancial aspects of marketplace transactions. One of James's passions is the many ways behavioral economics could play out in policy decisions, and over the years I have come to see the wisdom in this approach. James is currently a professor at the University of St. Thomas (in Minnesota, not the Virgin Islands).

Leonard Lee

Leonard joined the PhD program at MIT to work on topics related to e-commerce. Since we both kept long hours, we started taking breaks together late at night, and this gave us a chance to start working jointly on a few research projects. The collaboration with Leonard has been great. He has endless energy and enthusiasm, and the number of experiments he can carry out during an average week is about what other people do in a semester. In addition, he is one of the nicest people I have ever met and always a delight to chat and work with. Leonard is currently a professor at Columbia University.

Jonathan Levav

Jonathan loves his mother like no one else I have met, and his main regret in life is that he disappointed her when he didn't go to medical school. Jonathan is smart, funny, and an incredibly social animal, able to make new friends in fractions of seconds. He is physically big with a large head, large teeth, and an even larger heart. Jonathan is currently a professor at Columbia University.

George Loewenstein

George is one of my first, favorite, and longest-time collaborators. He is also my role model. In my mind George is the most creative and broadest researcher in behavioral economics. George has an incredible ability to observe the world around him and find nuances of behavior that are important for our understanding of human nature as well as for policy.

George is currently, and appropriately, the Herbert A. Simon Professor of Economics and Psychology at Carnegie Mellon University.

Nina Mazar

Nina first came to MIT for a few days to get feedback on her research and ended up staying for five years. During this time we had oodles of fun working together and I came to greatly rely on her. Nina is oblivious of obstacles, and her willingness to take on large challenges led us to carry out some particularly difficult experiments in rural India. For many years I hoped that she would never decide to leave; but, alas, at some point the time came: she is currently a professor at the University of Toronto. In an alternative reality, Nina is a high-fashion designer in Milan, Italy.

Elie Ofek

Elie is an electrical engineer by training who then saw the light (or so he believes) and switched to marketing. Not surprisingly, his main area of research and teaching is innovations and high-tech industries. Elie is a great guy to have coffee with because he has interesting insights and perspectives on every topic. Currently, Elie is a professor at Harvard Business School (or as its members call it, "The Haaarvard Business School").

Yesim Orhun

Yesim is a true delight in every way. She is funny, smart, and sarcastic. Regrettably, we had only one year to hang out while we were both at Berkeley. Yesim's research takes findings from behavioral economics and, using this starting point, provides prescriptions for firms and policy makers.

For some odd reason, what really gets her going is any research question that includes the words *simultaneity* and *endogeneity*. Yesim is currently a professor at the University of Chicago.

Drazen Prelec

Drazen is one of the smartest people I have ever met and one of the main reasons I joined MIT. I think of Drazen as academic royalty: he knows what he is doing, he is sure of himself, and everything he touches turns to gold. I was hoping that by osmosis, I would get some of his style and depth, but having my office next to his was not sufficient for this. Drazen is currently a professor at MIT.

Kristina Shampanier

Kristina came to MIT to be trained as an economist, and for some odd but wonderful reason elected to work with me. Kristina is exceptionally smart, and I learned a lot from her over the years. As a tribute to her wisdom, when she graduated from MIT, she opted for a nonacademic job: she is now a high-powered consultant in Boston.

Jiwoong Shin

Jiwoong is a yin and yang researcher. On one hand he carries out research in standard economics assuming that individuals are perfectly rational; on the other hand he carries out research in behavioral economics showing that people are irrational. He is thoughtful and reflective—a philosophical type—and this duality does not faze him. Jiwoong and I started working together mostly because we wanted to have fun together, and indeed we have spent many exciting hours working together. Jiwoong is currently a professor at Yale University.

Baba Shiv

Baba and I first met when we were both PhD students at Duke. Over the years Baba has carried out fascinating research in many areas of decision making, particularly on how emotions influence decision making. He is terrific in every way and the kind of person who makes everything around him seem magically better. Baba is currently a professor at Stanford University.

Rebecca Waber

Rebecca is one of the most energetic and happiest people I have ever met. She is also the only person I ever observed to burst out laughing while reading her marriage vows. Rebecca is particularly interested in research on decision making applied to medical decisions, and I count myself as very lucky that she chose to work with me on these topics. Rebecca is currently a graduate student at the Media Laboratory at MIT.

Klaus Wertenbroch

Klaus and I met when he was a professor at Duke and I was a PhD student. Klaus's interest in decision making is mostly based on his attempts to make sense of his own deviation from rationality, whether it is his smoking habit or his procrastination in delaying work for the pleasure of watching soccer on television. It was only fitting that we worked together on procrastination. Klaus is currently a professor at INSEAD.

Notes

1. James Choi, David Laibson, and Brigitte Madrian, "$100 Bills on the Sidewalk: Suboptimal Saving in 401(k) Plans," Yale University, working paper.
2. Steven Levitt and John List, "*Homo economicus* Evolves," *Science* (2008).
3. David Brooks, "The Behavioral Revolution," *New York Times* (October 27, 2008).
4. Jodi Kantor, "Entrees Reach $40," *New York Times* (October 21, 2006).
5. Itamar Simonson, "Get Closer to Your Customers by Understanding How They Make Choices," *California Management Review* (1993).
6. Louis Uchitelle, "Lure of Great Wealth Affects Career Choices," *New York Times* (November 27, 2006).
7. Katie Hafner, "In the Web World, Rich Now Envy the Superrich," *New York Times* (November 21, 2006).

8. Valerie Ulene, "Car Keys? Not So Fast," *Los Angeles Times* (January 8, 2007).

9. John Leland, "Debtors Search for Discipline through Blogs," *New York Times* (February 18, 2007).

10. Colin Schieman, "The History of Placebo Surgery," University of Calgary (March 2001).

11. Margaret Talbot, "The Placebo Prescription," *New York Times* (June 9, 2000).

12. Sarah Bakewell, "Cooking with Mummy," *Fortean Times* (July 1999).

13. D. J. Swank, S. C. G Swank-Bordewijk, W. C. J. Hop, et al., "Laparoscopic Adhesiolysis in Patients with Chronic Abdominal Pain: A Blinded Randomised Controlled Multi-Center Trial," *Lancet* (April 12, 2003).

14. "Off-Label Use of Prescription Drugs Should Be Regulated by the FDA," Harvard Law School, Legal Electronic Archive (December 11, 2006).

15. Irving Kirsch, "Antidepressants Proven to Work Only Slightly Better Than Placebo," *Prevention and Treatment* (June 1998).

16. Sheryl Stolberg, "Sham Surgery Returns as a Research Tool," *New York Times* (April 25, 1999).

17. Margaret E. O'Kane, National Committee for Quality Assurance, letter to the editor, *USA Today* (December 11, 2006).

18. Federal Bureau of Investigation, *Crime in the United States 2004—Uniform Crime Reports* (Washington, D.C.: U.S. Government Printing Office, 2005).

19. Brody Mullins, "No Free Lunch: New Ethics Rules Vex Capitol Hill," *Wall Street Journal* (January 29, 2007).

20. "Pessimism for the Future," *California Bar Journal* (November 1994).
21. Maryland Judicial Task Force on Professionalism (November 10, 2003): http://www.courts.state.md.us/publications/professionalism2003.pdf.
22. Florida Bar/Josephson Institute Study (1993).
23. *DPA Correlator*, Vol. 9, No. 3 (September 9, 2002). See also Steve Sonnenberg, "The Decline in Professionalism—A Threat to the Future of the American Association of Petroleum Geologists," *Explorer* (May 2004).
24. Jan Crosthwaite, "Moral Expertise: A Problem in the Professional Ethics of Professional Ethicists," *Bioethics*, Vol. 9 (1995): 361–379.
25. The 2002 Transparency International Corruption Perceptions Index, transparency.org.
26. McKinsey and Company, "Payments: Charting a Course to Profits" (December 2005).
27. "Email Has Made Slaves of Us," *Australian Daily Telegraph* (June 16, 2008).
28. "Studies Find Big Benefits in Marriage," *New York Times* (April 10, 1995).
29. Ralph Keeney, "Personal Decisions Are the Leading Cause of Death," *Operation Research* (2008).
30. "Pearls Before Breakfast," *Washington Post* (April 8, 2007).
31. John Maurice Clark, "Economics and Modern Psychology," *Journal of Political Economy* (1918).
32. http://www.openleft.com/showDiary.do?diaryId=8374, Openleft.com (posted September 21, 2008).
33. Dominique de Quervain, Urs Fischbacher, Valerie Treyer, Melanie Schellhammer, Ulrich Schnyder, Alfred Buck,

and Ernst Fehr, "The Neural Basis of Altruistic Punishment," *Science* (2004).

34. Martin Seligman and Steve Maier, "Failure to Escape Traumatic Shock," *Journal of Experimental Psychology* (1967).

35. Michael Crichton, *The Lost World* (New York: Random House, 1995).

Bibliography and Additional Readings

Below is a list of the papers on which the chapters were based, plus suggestions for additional readings on each topic.

Introduction

RELATED READINGS

Daniel Kahneman, Barbara L. Fredrickson, Charles A. Schreiber, and Donald A. Redelmeier, "When More Pain Is Preferred to Less: Adding a Better End," *Psychological Science* (1993).

Donald A. Redelmeier and Daniel Kahneman, "Patient's Memories of Painful Medical Treatments—Real-Time and Retrospective Evaluations of Two Minimally Invasive Procedures," *Pain* (1996).

Dan Ariely, "Combining Experiences over Time: The Effects of Duration, Intensity Changes, and On-Line Measurements on Retrospective Pain Evaluations," *Journal of Behavioral Decision Making* (1998).

Dan Ariely and Ziv Carmon, "Gestalt Characteristics of Experienced Profiles," *Journal of Behavioral Decision Making* (2000).

Chapter 1: The Truth about Relativity

RELATED READINGS

Amos Tversky, "Features of Similarity," *Psychological Review*, Vol. 84 (1977).

Amos Tversky and Daniel Kahneman, "The Framing of Decisions and the Psychology of Choice," *Science* (1981).

Joel Huber, John Payne, and Chris Puto, "Adding Asymmetrically Dominated Alternatives: Violations of Regularity and the Similarity Hypothesis," *Journal of Consumer Research* (1982).

Itamar Simonson, "Choice Based on Reasons: The Case of Attraction and Compromise Effects," *Journal of Consumer Research* (1989).

Amos Tversky and Itamar Simonson, "Context-Dependent Preferences," *Management Science* (1993).

Dan Ariely and Tom Wallsten, "Seeking Subjective Dominance in Multidimensional Space: An Explanation of the Asymmetric Dominance Effect," *Organizational Behavior and Human Decision Processes* (1995).

Constantine Sedikides, Dan Ariely, and Nils Olsen, "Contextual and Procedural Determinants of Partner Selection: On Asymmetric Dominance and Prominence," *Social Cognition* (1999).

Chapter 2: The Fallacy of Supply and Demand

BASED ON

Dan Ariely, George Loewenstein, and Drazen Prelec, "Coherent Arbitrariness: Stable Demand Curves without

Stable Preferences," *Quarterly Journal of Economics* (2003).

Dan Ariely, George Loewenstein, and Drazen Prelec, "Tom Sawyer and the Construction of Value," *Journal of Economic Behavior and Organization* (2006).

RELATED READINGS

Cass R. Sunstein, Daniel Kahneman, David Schkade, and Ilana Ritov, "Predictably Incoherent Judgments," *Stanford Law Review* (2002).

Uri Simonsohn, "New Yorkers Commute More Everywhere: Contrast Effects in the Field," *Review of Economics and Statistics* (2006).

Uri Simonsohn and George Loewenstein, "Mistake #37: The Impact of Previously Faced Prices on Housing Demand," *Economic Journal* (2006).

Chapter 3: The Cost of Zero Cost

BASED ON

Kristina Shampanier, Nina Mazar, and Dan Ariely, "How Small Is Zero Price? The True Value of Free Products," *Marketing Science* (2007).

RELATED READINGS

Daniel Kahneman and Amos Tversky, "Prospect Theory: An Analysis of Decision under Risk," *Econometrica* (1979).

Eldar Shafir, Itamar Simonson, and Amos Tversky, "Reason-Based Choice," *Cognition* (1993).

Chapter 4: The Cost of Social Norms

BASED ON

Uri Gneezy and Aldo Rustichini, "A Fine Is a Price," *Journal of Legal Studies* (2000).

James Heyman and Dan Ariely, "Effort for Payment: A Tale of Two Markets," *Psychological Science* (2004).

Kathleen Vohs, Nicole Mead, and Miranda Goode, "The Psychological Consequences of Money," *Science* (2006).

RELATED READINGS

Margaret S. Clark and Judson Mills, "Interpersonal Attraction in Exchange and Communal Relationships," *Journal of Personality and Social Psychology*, Vol. 37 (1979), 12–24.

Margaret S. Clark, "Record Keeping in Two Types of Relationships," *Journal of Personality and Social Psychology*, Vol. 47 (1984).

Alan Fiske, "The Four Elementary Forms of Sociality: Framework for a Unified Theory of Social Relations," *Psychological Review* (1992).

Pankaj Aggarwal, "The Effects of Brand Relationship Norms on Consumer Attitudes and Behavior," *Journal of Consumer Research* (2004).

Chapter 5: The Influence of Arousal

BASED ON

Dan Ariely and George Loewenstein, "The Heat of the Moment: The Effect of Sexual Arousal on Sexual Decision Making," *Journal of Behavioral Decision Making* (2006).

RELATED READINGS

George Loewenstein, "Out of Control: Visceral Influences on Behavior," *Organizational Behavior and Human Decision Processes* (1996).

Peter H. Ditto, David A. Pizarro, Eden B. Epstein, Jill A. Jacobson, and Tara K. McDonald, "Motivational Myopia: Visceral Influences on Risk Taking Behavior," *Journal of Behavioral Decision Making* (2006).

Chapter 6: The Problem of Procrastination and Self-Control
BASED ON

Dan Ariely and Klaus Wertenbroch, "Procrastination, Deadlines, and Performance: Self-Control by Precommitment," *Psychological Science* (2002).

RELATED READINGS

Ted O'Donoghue and Mathew Rabin, "Doing It Now or Later," *American Economic Review* (1999).

Yaacov Trope and Ayelet Fishbach, "Counteractive Self-Control in Overcoming Temptation," *Journal of Personality and Social Psychology* (2000).

Chapter 7: The High Price of Ownership
BASED ON

Ziv Carmon and Dan Ariely, "Focusing on the Forgone: How Value Can Appear So Different to Buyers and Sellers," *Journal of Consumer Research* (2000).

James Heyman, Yesim Orhun, and Dan Ariely, "Auction Fever: The Effect of Opponents and Quasi-Endowment on Product Valuations," *Journal of Interactive Marketing* (2004).

RELATED READINGS

Richard Thaler, "Toward a Positive Theory of Consumer Choice," *Journal of Economic Behavior and Organization* (1980).

Jack Knetsch, "The Endowment Effect and Evidence of Nonreversible Indifference Curves," *American Economic Review*, Vol. 79 (1989), 1277–1284.

Daniel Kahneman, Jack Knetsch, and Richard Thaler, "Experimental Tests of the Endowment Effect and the Coase Theorem," *Journal of Political Economy* (1990).

Daniel Kahneman, Jack Knetsch, and Richard H. Thaler, "Anomalies: The Endowment Effect, Loss Aversion, and Status Quo Bias," *Journal of Economic Perspectives*, Vol. 5 (1991), 193–206.

Chapter 8: Keeping Doors Open

BASED ON

Jiwoong Shin and Dan Ariely, "Keeping Doors Open: The Effect of Unavailability on Incentives to Keep Options Viable," *Management Science* (2004).

RELATED READINGS

Sheena Iyengar and Mark Lepper, "When Choice Is Demotivating: Can One Desire Too Much of a Good Thing?" *Journal of Personality and Social Psychology* (2000).

Daniel Gilbert and Jane Ebert, "Decisions and Revisions: The Affective Forecasting of Changeable Outcomes," *Journal of Personality and Social Psychology* (2002).

Ziv Carmon, Klaus Wertenbroch, and Marcel Zeelenberg, "Option Attachment: When Deliberating Makes Choosing Feel Like Losing," *Journal of Consumer Research* (2003).

Chapter 9: The Effect of Expectations

BASED ON

John Bargh, Mark Chen, and Lara Burrows, "Automaticity of Social Behavior: Direct Effects of Trait Construct and

Stereotype Activation on Action," *Journal of Personality and Social Psychology* (1996).

Margaret Shin, Todd Pittinsky, and Nalini Ambady, "Stereotype Susceptibility: Identity Salience and Shifts in Quantitative Performance," *Psychological Science* (1999).

Sam McClure, Jian Li, Damon Tomlin, Kim Cypert, Latané Montague, and Read Montague, "Neural Correlates of Behavioral Preference for Culturally Familiar Drinks," *Neuron* (2004).

Leonard Lee, Shane Frederick, and Dan Ariely, "Try It, You'll Like It: The Influence of Expectation, Consumption, and Revelation on Preferences for Beer," *Psychological Science* (2006).

Marco Bertini, Elie Ofek, and Dan Ariely, "To Add or Not to Add? The Effects of Add-Ons on Product Evaluation," Working Paper, HBS (2007).

RELATED READINGS

George Loewenstein, "Anticipation and the Valuation of Delayed Consumption," *Economic Journal* (1987).

Greg Berns, Jonathan Chappelow, Milos Cekic, Cary Zink, Giuseppe Pagnoni, and Megan Martin-Skurski, "Neurobiological Substrates of Dread," *Science* (2006).

Chapter 10: The Power of Price

BASED ON

Leonard Cobb, George Thomas, David Dillard, Alvin Merendino, and Robert Bruce, "An Evaluation of Internal Mammary Artery Ligation by a Double-Blind Technic," *New England Journal of Medicine* (1959).

Bruce Moseley, Kimberly O'Malley, Nancy Petersen, Terri Menke, Baruch Brody, David Kuykendall, John Hollingsworth,

Carol Ashton, and Nelda Wray, "A Controlled Trial of Arthroscopic Surgery for Osteoarthritis of the Knee," *New England Journal of Medicine* (2002).

Baba Shiv, Ziv Carmon, and Dan Ariely, "Placebo Effects of Marketing Actions: Consumers May Get What They Pay For," *Journal of Marketing Research* (2005).

Rebecca Waber, Baba Shiv, Ziv Carmon, and Dan Ariely, "Commercial Features of Placebo and Therapeutic Efficacy," *JAMA* (2008).

RELATED READINGS

Tor Wager, James Rilling, Edward Smith, Alex Sokolik, Kenneth Casey, Richard Davidson, Stephen Kosslyn, Robert Rose, and Jonathan Cohen, "Placebo-Induced Changes in fMRI in the Anticipation and Experience of Pain," *Science* (2004).

Alia Crum and Ellen Langer, "Mind-Set Matters: Exercise and the Placebo Effect," *Psychological Science* (2007).

Chapters 11 and 12:
The Context of Our Character, Parts I and II

BASED ON

Nina Mazar and Dan Ariely, "Dishonesty in Everyday Life and Its Policy Implications," *Journal of Public Policy and Marketing* (2006).

Nina Mazar, On Amir, and Dan Ariely, "The Dishonesty of Honest People: A Theory of Self-Concept Maintenance," *Journal of Marketing Research* (2008).

RELATED READINGS

Max Bazerman and George Loewenstein, "Taking the Bias out of Bean Counting," *Harvard Business Review* (2001).

Max Bazerman, George Loewenstein, and Don Moore,

"Why Good Accountants Do Bad Audits: The Real Problem Isn't Conscious Corruption. It's Unconscious Bias," *Harvard Business Review* (2002).

Maurice Schweitzer and Chris Hsee, "Stretching the Truth: Elastic Justification and Motivated Communication of Uncertain Information," *Journal of Risk and Uncertainty* (2002).

Chapter 13: Beer and Free Lunches

BASED ON

Dan Ariely and Jonathan Levav, "Sequential Choice in Group Settings: Taking the Road Less Traveled and Less Enjoyed," *Journal of Consumer Research* (2000).

Richard Thaler and Shlomo Benartzi, "Save More Tomorrow: Using Behavioral Economics to Increase Employee Savings," *Journal of Political Economy* (2004).

RELATED READINGS

Eric J. Johnson and Daniel Goldstein, "Do Defaults Save Lives?" *Science*, (2003).

Index

Index

Index

Index

Dunkin' Donuts, moving anchor to Starbucks from, 37–39

DVD players, FREE! DVD offers and, 55

E

earmarking, congressional restrictions on, 204–5

Ebbers, Bernie, 223

economics, standard:
arbitrary coherence at odds with, 43, 45, 47–48
behavioral economics vs., xxviii–xxx, 239–40
cost-benefit analysis in, 64–65
human rationality assumed in, xxix, xxx, 239–40
supply and demand in, 45–46

Economist subscription offers, 1–3, 4–6, 9–10

education, 84–86
igniting social passion for, 85–86
"No Child Left Behind" policy and, 85

"elderly," behavior affected by priming concept of, 170–71

e-mail addiction, 255–59
overcoming, 259
reinforcement schedules and, 257–59

empirical tests:
public policy and, 328–29
in science, xxv–xxvi, 325

employees:
payment of, *see* compensation; salaries
social vs. market norms in companies' relations with, 80–84, 252–54
theft and fraud at workplace ascribed to, 195–96

endowment effect, 129–35

energy drinks, impact of price and hype on efficacy of, 184–87

Enron scandal, xiv, 196, 204, 219

envy, comparisons and, 15–19

epidurals, 103–4

Escape from Freedom (Fromm), 148

Europe, savings rate in, 109

evolution, dangers of globalization and, 317–18, 319

exercise, procrastination and, 111

expectations, 155–72, 269–75
art and, 274
beer experiments and, 157–59, 161–62, 163–64, 172

brand associations of Coke and Pepsi and, 166–68

conflicts and, 156–57, 171–72

depth of description in caterers' offerings and, 164

exotic-sounding ingredients and, 164–65

football plays and, 155–56, 171

garage sales and, 162–63

knowledge before vs. after experience and, 161–64

marketing hype and, 186–87

music and, 270–73, 274

physiology of experience altered by, 161–64, 166–68, 293–94

placebo effect and, 173–94; *see also* placebo effect

restaurant meals and, 269–70

sports car test drives and, 161

stereotypes and, 168–71

taste and, 157–68, 270

upscale coffee ambience and, 159–60

wineglasses and, 165

expense reports, dishonesty in, 223–24

experience, not learning from, xxvii

experiments:
extrapolation of findings in, xxxi–xxxii
isolating individual forces in, xxxi
see also empirical tests; *specific topics*

F

Fannie Mae, 280, 310

Fastow, Andrew, 219

Federal Depositor Insurance Corporation (FDIC), 280

Federal Reserve, 280, 284–85

Fehr, Ernst, 307–8

financial industry:
conflicts of interest and, 295–96
globalization and loss of diversity in, 318–19
inherent fuzziness in, 294–95
profit made from our mistakes by, 298–304
regulation of, 296
see also bankers

financial meltdown of 2008, 279–329
bailout plan and, 280, 304–6, 310–11, 312, 314, 319–20
bankers' behavior in, 291–96
collapse of financial institutions in, 280–81, 314

Index

financial meltdown of 2008 (*continued*)
  compensation for bankers and, 306,
    310, 311, 319–24
  conflicts of interest and, 291–96
  empirical testing of approaches to,
    328–29
  global market and, 316–19
  Greenspan's confession and, xvii-xix
  housing market collapse and, 265–66,
    279
  learned helplessness and, 314–16
  limitations of rational economics and,
    281–82, 324–28
  media coverage of, 315–16
  mortgage practices and, 279–80,
    283–90
  planning fallacy and, 297–304
  psychological fallout from not
    understanding what's going on in,
    311–16
  public trust and, 304–11
  shared suffering in, 303
fines, in social context, 76–77
first decisions:
  power of, 44
  shape of our lives and, 43
  translation of, into long-term habits,
    36–39
  see also anchoring
first impressions:
  imprinting and, 25, 34, 43
  see also arbitrary coherence
Fiske, Alan, 68
food:
  expectations and taste of, 164–65, 270
  ordering process and enjoyment of,
    237–38
  see also taste
food labels, allure of "zero" on, 61–62
football plays, expectations and
  perception of, 155–56, 171
Ford Motor Company, 119–21
401(k)s, xiii
France, Amazon's FREE! shipping
  promotion in, 59, 62
Freddie Mac, 280, 310
Frederick, Shane, 157, 161, 336
FREE!, 49–63, 247–50
  Amazon gift certificate offer and, 58
  AOL price structure and, 59–60
  checking accounts or mortgages and,
    60, 301–2

chocolate pricing experiment and,
  51–54, 64–65
credit cards and, 247–48
exchanges and, 55–58
fear of loss and, 54–55
high-definition DVD players and, 55
history of zero and, 50
museum admission fees and, 61
oil changes with car purchases and,
  60–61
preventive health care and, 62–63
rational cost-benefit analysis and, 64–65
restaurant meals with friends and,
  248–50
shipping offers on orders over a certain
  amount and, 58–59, 62
social policy and, 62–63
time considerations and, 61
free, working for, 71
free lunches, 240–44
free market, 47–48
free trade, 47–48
Frenk, Hanan, xxv
frequent-flyer miles, 227–28
Freud, Sigmund, 98, 203
friendly requests, social norms and, 68,
  70–71, 73–74, 77–78
Fromm, Erich, 148
functional magnetic resonance imaging
  (fMRI), taste test of Coke and Pepsi
  and, 166–68
furniture, assembling, pride of ownership
  and, 135
fuzziness, in financial world, 294–95

G
gambling, 258
garage sales, 129–30, 162–63
gasoline, price increases and demand for,
  47
Gell-Mann, Murray, 244
gender stereotypes, 169
Gerbi (Italian physician), 177
gift certificate experiment, 58
gifts:
  Burning Man based on exchange of,
    86–88
  cash vs., as employee reward, 82–83,
    253–54
  mere mention of money and, 73–74
  in social situations, 250–52
  social vs. market norms and, 72–74

360

Index

globalization, 316–19
  diversity, competition, and efficiency
    lessened by, 318–19
Gneezy, Uri, 76–77, 320–21
Gone with the Wind (Mitchell), 150
Goode, Miranda, 74–75
Google, 83
goslings, imprinting in, 25, 34, 43
government:
  social contract between citizens and, 84
  see also Congress, U.S.
Greenspan, Alan, xvii–xix
gridlock, legislative, 151, 152
*Guidelines for Lawyer Courtroom
  Conduct* (Sweeney), 213
guilt, social norms and, 77

H
habits:
  first decisions translated into, 36–38
  questioning, 44
Halloween experiment, 56–58
Hamlet (Shakespeare), xxviii–xxix, 232
Harford, Tim, 291–92
Harvard Business School, 197–98
  honesty experiment at, 198–202
health care, 110–11
  bundling of medical tests and
    procedures and, 119–21
  conflicts of interest in, 293, 295
  defeating procrastination in, 117–21
  FREE! procedures and, 62–63
  mandatory checkups and, 118
  patient compliance and, 260–64
  placebo effect and, 173–94, 275–78; *see
    also* placebo effect
  price of medical treatments and, 176,
    180–87, 190
  public policy and spending on, 190
  scientifically controlled trials and, 173–76
  self-imposed deadlines and, 118–19
helping, thinking about money and, 74, 75
herding, 36–38
  self-herding and, 37–38
Heyman, James, 69–71, 136, 336–37
HIV-AIDS, 90
Holy Roman emperors, placebo effect
  and, 188
Home Depot, 78
Honda, 120, 121
honesty, 195–230
  contemplation of moral benchmarks
    and, 206–9, 213

dealing with cash and, 217–30
importance of, 214–15
as moral virtue, 203
oaths and, 208–9, 211–13, 215
reward centers in brain and, 203, 208
Smith's explanation for, 202, 214
superego and, 203–4, 208
  see also dishonesty
Hong, James, 21
honor codes, 212–13
hormones, expectation and, 179
house sales:
  anchoring and, 30-31
  relativity and, 8–9, 19
  value in owner's eyes and, 129, 135,
    265–69
housing market:
  bubble in, 289–90
  decreasing valuations and, 265–66, 279

I
ice cream, FREE!, time spent on line for, 61
"Ikea effect," 135
immediate gratification:
  e-mail and, 255–59
  unpleasant medical treatments and,
    261–64
imprinting, 25, 34, 43
  see also anchoring
indecision, 151–53
individualism, 68
  thinking about money and, 74, 75
ingredients, exotic-sounding, 164–65
innovation, increased globalization and,
  316–18
insurance fraud, 196, 223
insurance industry, 296
  punitive finance practices of, 299-301
  spreading cost of, 304
interest-only mortgages, 287–88
interferon, 260–64
internal mammary artery ligation,
  173–74, 191
inventiveness, 68
IRA (Irish Republican Army),
  156–57
Iran, lack of trust in, 214–15
irrational behaviors, xxix–xxx
  opportunities for improvement and,
    240–44
  systematic and predictable nature of,
    xxx, 239
  see also specific topics

Index

Index

Mead, Nicole, 74–75
medical benefits, recent cuts in, 82
medical care, *see* health care
medical profession:
  conflicts of interest and, 293, 295
  decline of ethics and values in, 210
  salaries of, as practicing physicians vs.
    Wall Street advisers, 18–19
memory of previous prices, price changes
  and, 46–47
Mencken, H. L., 17–18
menu pricing, in restaurants, 4
Merrill Lynch, 280, 310
Mills, Judson, 68
mistakes, repeated, and failure to learn
  from experience, xxvii
Mitchell, Margaret, 150
MIT Sloan School of Management, 92
Mona Lisa (Leonardo), 274
money:
  benefits of, 86
  dishonesty with nonmonetary objects
    vs., 217–30
  doing away with, 86–88
  impact of mere mention of, 73–75
  switch away from, to electronic
    instruments, 230
Montague, Latané, 166–68
Montague, Read, 166–68
moral benchmarks, dishonesty curbed by
  contemplation of, 206–9, 213
morality:
  in "cold" vs. aroused state, 94–95, 96, 97
  see also cheating on tests; dishonesty;
    honesty
mortgage-backed securities, 279–80,
  294–95, 305
mortgages, 60
  calculating optimal amount of, 283–90
  government regulation of, 290
  interest-only, 287–88
  subprime, 303
Moseley, J. B., 174–76
movie reviews, enjoyment affected by, 166
mummy powder, 177–78
museums, free-entrance days or times at,
  61
music, expectations and, 270–73, 274

N

need for uniqueness, ordering food or
  drinks and, 237–38

New England Journal of Medicine, 175
news reports, on financial meltdown,
  315–16
New York Times, 4, 18, 21, 122–23
Niskanen, William A., 205–6
"No Child Left Behind" policy, 85
Norton, Mike, 135
nucleus accumbens, 203, 208

O

oaths, honesty and, 208–9, 211–13, 215
Obama, Barack, 323–24
Ofek, Elie, 159–60, 338
online auctions, 135–36
open-source software, 81
options, 139–53
  abundance of, in modern democracy,
    148
  aversion to loss and, 148–49
  college students' choice of major and,
    141–42
  consciously closing, 150–51
  "door game" and, 143–48
  downside of, 140
  important, vanishing of, 149
  romantic relationships and, 142, 148, 150
  sale prices and, 148–49
  similar, choosing between, 151–53
  Xiang Yu's story and, 139–40
ordering food or drinks, 231–38
  enjoyment of choices and, 232, 235–36,
    237, 238
  need for uniqueness and, 237–38
  out loud vs. in private, 231–32, 233–36,
    237–38
  strategy for, 238
Orhun, Yesim, 136, 338–39
osteoarthritis, arthroscopic knee surgery
  and, 174–76
outsourcing, 81–82
overdraft fees, 301–2
ownership, 127–38
  aversion to loss and, 134, 137, 138
  Duke University basketball tickets and,
    127–33
  of points of view, 137–38
  pride of, putting work into something
    and, 135
  "trial" promotions and, 136–37
  value in owner's eyes increased by,
    129–35, 265–69
  virtual, online auctions and, 135–36

Index

productivity, social norms in workplace and, 80–84

profession, origin of word, 209

professional ethics, 215
 attempts at improvement of, 213–14
 decline in, 209–11

professional oaths, 208–9, 213

punishment:
 immediate, desired behavior resulting in, 263
 unpredictable, learned helplessness and, 312–16

purchases, price imprinting and, 30

Q

Qin (Ch'in) dynasty, 139

R

Rapp, Gregg, 4

rational economic model, xii–xvii, xxix, xxx, 239–40, 279, 281–82, 307, 318, 324–29

reciprocity, social vs. market norms and, 68–69

regulation, 325
 of banking and financial industries, 296
 mortgage limits and, 290
 self-destructive behaviors restrained by, 118, 290

reinforcement schedules, fixed vs. variable, 257–59
 overcoming e-mail addiction and, 259

relativity, 1–21
 bread-making machines and, 14–15
 changing focus from narrow to wide and, 19–20
 compensation of bankers and, 324
 controlling circles of comparison and, 19, 21
 dating and, 10–14, 15, 245–46
 dealing with problem of, 19–21
 decoy effect and, 5–6, 8–15, 245–46
 Economist subscription offers and, 1–3, 4–6, 9–10
 house purchases and, 8–9, 19
 jealousy and envy springing from, 15–19
 prices for various products and, 29
 restaurant menu pricing and, 4
 salaries and, 16–19
 television pricing and, 3–4
 tendency to compare things that are easily comparable and, 8–9

traveling and, 246–47
 vacation planning and, 10
 visual demonstration of, 7

relocation, anchoring to housing prices and, 30–31

restaurants:
 expectations and, 269–70
 FREE! approach to dining with friends in, 248–50
 with lines to get in, 36, 37
 menu pricing of, 4
 ordering in, 231–38; *see also* ordering food or drinks
 social norms of dating and, 75–76

retirement, saving for, xiii
 from perspective of standard economics vs. behavioral economics, 241

revenge, 307–9
 as enforcement mechanism, 309
 pleasure of, 307–8

rewards:
 delayed gratification and, 261–64
 financial, job performance and, 320–21
 reinforcement schedules and, 257–58
 unpredictable, learned helplessness and, 312–16
 see also bonuses

robberies, 195

romantic relationships:
 options in, 142, 148, 150
 separation of social and market norms and, 69, 75–76, 250
 see also dating

Roth, Al, xviii

royal touch, 188

Rustichini, Aldo, 76–77

S

safe sex, 100–102
 and willingness to engage in unprotected sex when aroused, 89, 95, 96–97, 99, 107

salaries, 16–19, 88
 of bankers, calculating right amount of, 319–24
 of CEOs, 16–17, 18, 310
 co-workers' comparisons of, 16
 excessive, erosion of public trust and, 310, 311
 executive, public outcry over, 306, 310, 311, 319–20
 happiness and, 17–18

Index

Index